AF809930

Routledge Revivals

Waterfowl and Wetlands

Originally published in 1974, *Waterfowl and Wetlands* analyses waterfowl hunting patterns in the late sixties in the hopes of protecting waterfowl resources such as wetlands. Wetlands are obviously an important resource for migratory waterfowl however they are often drained for agricultural purposes which can have dramatic effects on waterfowl population. This study aims to explore the issues surrounding waterfowl and wetlands in an attempt to determine their value to hunters, farmers and the general public. This study will be of interest to students of Environmental Studies.

percent decrease in the number of potholes would not reduce any species of waterfowl in North America to the endangered level, although the populations of several species would be markedly lower. He maintains that even at reduced levels sufficient birds of all species would remain and that, if well distributed, they would be capable of meeting a greatly increased demand for outdoor recreation in the form of bird watching or photography. These observations led him to conclude that if the objective is producing and distributing a waterfowl population to supply bird watchers, it is not likely that millions of dollars need be spent in preserving prairie wetland habitat.[3]

The point he makes is this: nonconsumptive uses of waterfowl exist and have economic value that is part of the total worth of a given waterfowl population. The value of nonconsumptive use is presumed to be insensitive to changes in the total size of the population and is therefore unaltered by relevant changes in the supply of nesting habitat. But the total value of consumptive use does vary with the size of the waterfowl population. Hence justification for habitat preservation or increase requires knowledge of the value additional waterfowl will have primarily for hunters.[4]

DIRECTION OF THE STUDY

The next chapter deals with alternative methods of valuing outdoor recreational resources in general, and develops a theoretical framework for the valuation of waterfowl in particular. Chapter 3 supplies empirical results derived from the valuation model. Turning from hunter demand to waterfowl supply, chapter 4 posits biometric relationships and tests them with mallard data. Chapter 5 uses information from earlier chapters to provide crude cost-benefit analyses on existing prairie wetlands and explores the questions of how many waterfowl and how many ponds are appropriate in a dynamic setting. The last chapter is a potpourri containing discussions of various aspects of this work, including limitations, conclusions, and recommendations for future study.

[3] Crissey, "Prairie Potholes," p. 171.

[4] Of the 148 million persons in the United States, 12 years old or older in 1965, approximately 22 percent fished, hunted, or did both for 3 days or more during the year. About 9 percent engaged in hunting, and 1 percent specifically engaged in the hunting of waterfowl. Statistics derived from data in U.S. Bureau of Census, *Statistical Abstract of the United States: 1968*, pp. 8, 204, and in M. Edwin Rosasco, David R. Anderson, and Elwood M. Martin, "Waterfowl Harvest and Hunter Activity in the United States During the 1966 Hunting Season," Administrative Report No. 138, Migratory Bird Populations Station (Laurel, Md.: U.S. Bureau of Sport Fisheries and Wildlife, 1967), p. 25.

chapter two Valuation of Waterfowl: A Theoretical Framework

Valuation of fish and game is an important segment of the young and imperfect field of recreational resource economics. The valuing of any recreational resource poses a formidable problem for economic analysis because most recreational resources are not priced in a market or are only partially priced through charges that seek to match supply with demand. (Campground fees and hunting license fees are examples of partial pricing.) Hence the economist's usual tools for valuation of goods and services, which are based on market observations, either are of little use or require considerable modification. No fully satisfactory general analytic framework has yet been developed, but continued attempts at analysis are spurred by the lack of value figures for rational decision making and increasing alternate-use values for many recreation sites.

There are valuation methods developed by others that merit comparative discussion both in general terms and with specific respect to waterfowl. Since most of these methods have consumer's surplus as their implicit or explicit foundation, this concept is developed here.

CONSUMER'S SURPLUS

Informally stated, consumer's surplus is a measure of net benefits—benefits above costs paid—received by an individual from some good or service

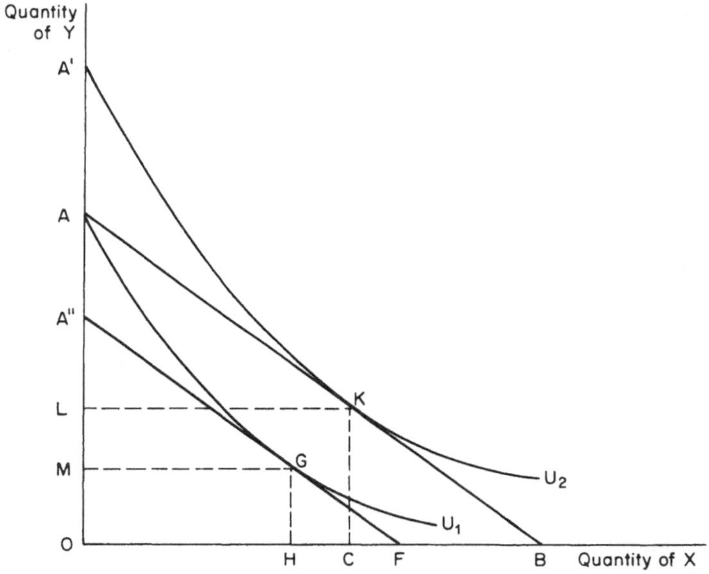

Figure 1. Indifference curves for an individual.

purchased. In more formal terms, consumer's surplus is a measure of welfare difference for an individual confronted with changed trading circumstances for a good.[1] There are many consumer's surplus measures, but discussion will be confined to two measures relevant to this analysis: willingness to pay and willingness to sell.

In figure 1, an individual has an initial endowment of income, Y, represented by OA; he is on indifference curve U_1.[2] He is then allowed the opportunity to purchase good X at the price represented by line AB.[3] To obtain the highest possible level of welfare, the individual trades until he

[1] Although the concept of consumer's surplus must first be developed in terms of the individual, interest usually lies instead in the summed consumers' surplus of a group of individuals. Using such a measure of welfare implicitly assumes that a dollar has equal "value," or utility, for each individual. The arbitrariness of this assumption demonstrates that consumers' surplus for a group is a rough approximation of welfare at best.

[2] An indifference curve represents various combinations of two goods which give the individual identical amounts of satisfaction. A higher indifference curve therefore represents a higher level of welfare. Figure 1 has a field of indifference curves, of which only two are shown. They are curved on the assumption that giving up a bit of something hurts more when one has relatively little of it.

[3] The price is indicated by the slope of the line, which is the price of X divided by the price of Y. The price of Y is, of course, equal to \$1. Since "price lines" AB and $A''F$ are parallel and have the same slope, they indicate a single price of X.

reaches point K on indifference curve U_2. He then has for consumption OC of good X and OL of all other goods as represented by income, Y. Two questions now may be framed; their answers will provide two measures of consumer's surplus:

1. Assuming the individual has reached point K, what is the smallest amount we must pay him to forgo consumption of X and yet leave him as well off as when he could buy X? The answer is AA', where the individual restricts his consumption to Y alone yet remains on indifference curve U_2, thus maintaining the same welfare level. This measure of consumer's surplus is the Hicksian "equivalent variation,"[4] or willingness to sell.

2. Assuming the individual is allowed to purchase X at the price represented by the slope of line AB, what is the maximum amount of income we may take away from him and yet leave him as well off as he was at A when he could not buy X at all? We may take away AA'' of Y, leaving the individual at point G on indifference curve U_1, consuming OH of X and OM of Y.

This measure of consumer's surplus, the Hicksian "compensating variation,"[5] or willingness to pay, may be represented as an area under a demand curve, as in figure 2. The income-compensated demand curve of our same individual for good X is shown by line NR. The price of X in figure 1, represented by the slope of price line AB, is represented in figure 2 by the line distance $O'P$. The individual buys $O'H'$ (equal to quantity OH, figure 1) of X, paying a total dollar amount for it represented by area $O'PSH'$. He obtains benefits from the purchase of X that are greater than the cost. These net benefits (consumer's surplus) are measured in dollars and are represented by area PNS, which is identical to the compensating variation represented in figure 1 by line length AA''.[6]

Thus two measures of consumer's surplus have been developed. Willingness to sell shows the *minimum* amount the individual must be paid to induce him to refrain from purchase and consumption of X; willingness to pay shows the *maximum* amount the individual would be willing to pay for the privilege of purchasing and consuming X at the specified price.

[4] J. R. Hicks, *Value and Capital*, 2nd ed. (Oxford: Clarendon Press, 1946), pp. 38–41, 329–333.

[5] *Ibid.*

[6] The demand curve in figure 2 also shows that the individual obtains fewer and fewer gross benefits from each additional unit of X purchased. The gross benefits per additional unit are represented in dollar terms by the line length from the demand curve to the x-axis, perpendicular to the latter at the point representing the appropriate quantity of X. The individual buys X until his gross benefits from the last (or marginal) unit of X are equal to the price of X; this maximizes his net benefits.

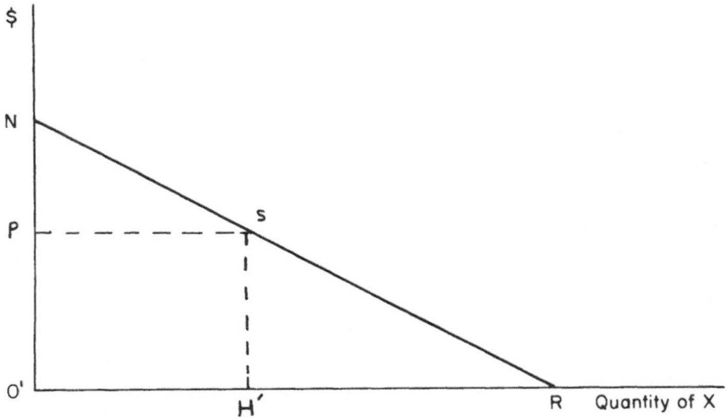

Figure 2. Demand curve for the same individual.

(The individual would be unwilling to pay an amount even infinitesimally greater than *AA″* in figure 1, for such payment would preclude maintenance of the level of welfare represented by indifference curve U_1.)

It was noted some years ago that usually only one measure of consumer's surplus is relevant to a specific situation.[7] The determination of the appropriate measure essentially involves the explicit specification of property rights, or, in weaker cases, implicit rights through prior use patterns. For example, if some alternate and incompatible use were to be considered for a presently existing state park, it would be appropriate to determine the recreation value of the park through individuals' willingness to sell their rights to use the facility for recreation. On the other hand, if a new state park were contemplated, willingness of individuals to pay for recreational use would measure the value of land now being used in some other fashion. We are, of course, assuming that the distribution of wealth is accepted. Wealth, in turn, is broadly interpreted to include the value of a reasonable expectation that land management policies will not be altered abruptly.

Generally speaking, even in the case of a single individual concerned with a single existing recreation site, valuation under willingness to sell will differ from valuation under willingness to pay. In common sense terms, a person cannot pay more than his discretionary income for continued use of the site, but he can require compensation greater than this amount to give up use voluntarily. In more technical terms, if the income elasticity

[7] A. Henderson, "Consumer's Surplus and the Compensating Variation," *The Review of Economic Studies*, vol. 8 (February 1941), pp. 117–121.

of the good is greater than zero,[8] the measure of willingness to sell will exceed that of willingness to pay. Only when the income elasticity is precisely zero will the two measures of consumer's surplus be equal.[9]

LITERATURE ON RECREATION RESOURCE VALUATION

The Davis Method

A review of selected literature on recreation valuation shows that the method making the most direct use of consumer's surplus is that developed by Robert K. Davis in a pioneering study of recreation benefits in the Maine woods.[10] Using a combination of interview and questionnaire techniques, data on willingness to pay and other variables were obtained from approximately 100 independent observations and on-site interviews with persons using the Maine woods for recreation. The willingness to pay was developed through "a bidding 'game' in which respondents could react positively, negatively or indifferently to increased costs of visiting the area. Bids were systematically raised (or lowered) until the user switched his reaction from inclusion to exclusion (or vice versa)."[11] Because of the remoteness of the geographic area, Davis assumed that each recreationist could use the area once annually; as a result of this indivisibility the first visit could be considered the marginal visit and the entire willingness to pay was therefore a marginal valuation. Willingness to pay was regressed on income, years of experience with the area, and length of visit, where the experience variable serves as a measure of taste and the visit length is considered a measure of the quality of the recreational experience rather than the quantity. Data on the independent variables for a considerably larger sample of the user population were obtained through a questionnaire. From the information derived, net benefits and simulated demand curves were calculated for three separate Maine forest areas.

Although the specific application used willingness to pay, the Davis method obviously can be adapted to a bidding game (or a questionnaire)

[8] Income elasticity is the percentage change of the quantity of a good purchased divided by the percentage change in income (which caused the quantity of the good purchased to change in the first place).

[9] Don Patinkin, "Demand Curves and Consumer's Surplus," in Carl F. Christ *et al.*, *Measurement in Economics* (Stanford,Calif.: Stanford University Press, 1963), pp. 94–95.

[10] "The Value of Outdoor Recreation: An Economic Study of the Maine Woods," unpublished Ph.D. thesis, Harvard University, 1963. Also, Robert K. Davis, "The Value of Big Game Hunting in a Private Forest," *Transactions of the Twenty-Ninth North American Wildlife and Natural Resources Conference* (Washington, D.C.: Wildlife Management Institute, 1964), pp. 393–403.

[11] *Ibid.*, p. 1 of summary.

eliciting willingness to sell; the general technique is clearly compatible with both measures of consumer's surplus, and rests easily upon the theoretical foundation. Practical objections can be raised, however, and are treated later.

Travel Cost and Transfer Cost Methods

Probably the oldest and most frequently used valuation method that warrants extended comment was suggested by Harold Hotelling and developed and refined by Marion Clawson and Jack L. Knetsch.[12] The technique is based on transportation and other costs of travel. Concentric rings or zones around the site are specified, and average travel costs and participation rates are determined in each of the geographic areas. This information can then be used to simulate a demand curve for a site over a range of user charges, and to estimate consumer's surplus. An example may help the reader grasp the essence of the technique: if recreationist A, representative of those who come from an outer zone, visits the site twice a year when his transportation cost is $10 per trip, and recreationist B, representative of those who come from an inner zone, visits the site five times a year when his transportation cost is $3, then it is concluded that a fee—say an entrance fee to the recreation site—of $7 per trip imposed upon recreationist B would decrease his trips from five to two. Hence recreation use by persons in more distant zones can be utilized to predict participation of those in inner zones under user charges, or prices, of various amounts. (To rid ourselves of a quantity unit problem, we assume throughout this section that each recreation visit consists of only one day at the recreation site.)

The elementary formulation of the technique requires the assumption of similar zonal population characteristics, i.e., similar distributions of income and tastes in each and every zone. More sophisticated treatment adjusts for differences in incomes between zones, but no study of this type to our knowledge has yet accounted for differences in tastes. The omission may be a serious one, for as Anthony Scott has pointed out, recreation sites are often adjacent to rural areas, with population centers

[12] Marion Clawson and Jack L. Knetsch, *Economics of Outdoor Recreation* (Baltimore: Johns Hopkins Press, for Resources for the Future, 1966). See also Marion Clawson, *Methods of Measuring the Demand for and Value of Outdoor Recreation*, Resources for the Future Reprint No. 10, February 1959. Harold Hotelling's letter to the National Park Service suggesting the general method is reproduced (without specific date, but written in 1949 or earlier) in William G. Brown, Amjer Singh, and Emery N. Castle, *An Economic Evaluation of the Oregon Salmon and Steelhead Sport Fishery*, Technical Bulletin No. 78 (Corvallis: Oregon State University Agricultural Experiment Station, 1964), pp. 6–7.

farther away, and the rural population may well differ from their city brethren in relevant taste factors.[13]

Even in the earliest stages, the travel cost approach tolerated considerable ambiguity regarding exactly which cost categories were to be included in analysis. Hotelling clearly intended only inclusion of transportation costs, while examples of the technique provided by Clawson and Knetsch appear to vary in the cost categories used. Historically, the literature has increased the number of categories considered appropriate for inclusion, often with considerable casualness and little attempt at theoretical justification.

An extreme example of cost inclusion is provided by William G. Brown, Amjer Singh, and Emery N. Castle in a study using transfer costs, "costs incurred by the buyer or seller of goods, but which are not normally included in prices."[14] Needless to say, such costs and cost categories are legion when considering the field of recreation, even when durable goods are excluded, as they were in the Brown, Singh, and Castle study.[15]

To shed some light on which costs should be included in a properly formulated analysis, let us recall that the technique is based on cost *differentials* between recreationists, and on the assumption that individuals incurring lesser costs would be willing to pay the cost differential as a fee. They would then adopt use patterns similar to those of individuals with greater costs. Since the technique is dependent upon cost differentials, the inclusion or exclusion of costs common to all in terms of the recreation quantity unit, such as daily expenditures by waterfowl hunters for shotgun shells or by fishermen for bait, is irrelevant and will be washed out in the empirical analysis unless the researcher commits an econometric error. The theoretically appropriate analysis would obtain a net economic value consistent with a measure of consumer's surplus, and since expenditures are being dealt with, the measure of relevance is willingness to pay. The analysis will have greatest theoretical justification if the following two criteria are met:

1. Any cost differential, or portion thereof included, has *no* satisfaction directly connected to it, and any cost differential, or portion thereof excluded, *does* have satisfaction equal to the cost directly connected to it.

2. Adequate adjustment is made for differences among individuals in income, tastes, and the like.

Let us first support our position that inclusion or exclusion of common

[13] "The Valuation of Game Resources: Some Theoretical Aspects," *Canadian Fisheries Reports*, No. 4 (May 1965), pp. 27–47.

[14] Brown *et al.*, *Economic Evaluation*, p. 10.

[15] *Ibid.*, pp. 18–23.

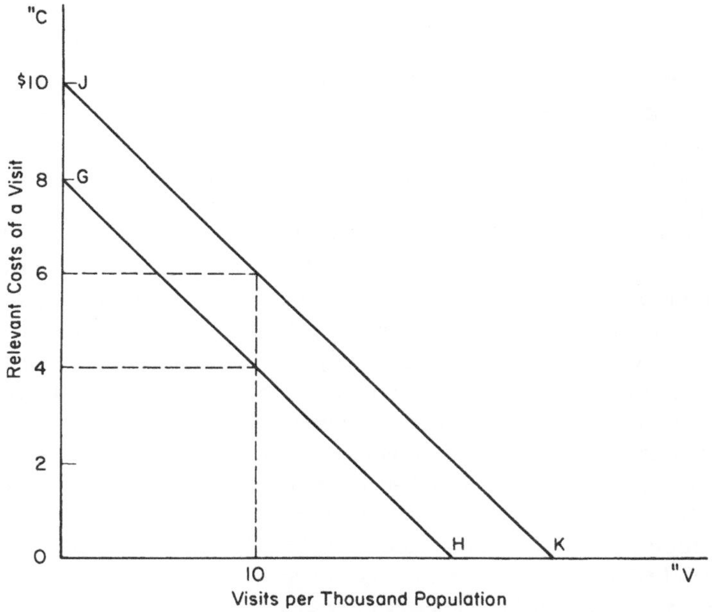

Figure 3. Cost–visitation relationships derived from zonal data.

costs is irrelevant, and do so in terms of the apparatus actually used in the travel and transfer cost approaches. The methods use a regression relating visits per thousand of the population from each zone (*"V*), to relevant costs of a visit from each zone (*"C*). From this relationship, total visits from all zones may be projected for various site entrance fees (prices), and these price–quantity relationships yield the aggregate demand curve for the site.[16]

For a waterfowl hunting site, assume the regression is in linear form, or

$$''V = ''a - ''b''C \tag{2-1}$$

and also assume the cost figures used exclude shotgun shell costs of $2 per visit common to all. The hypothetical regression results are illustrated by line *GH* in figure 3. If the regression is run again, using the same values for *"V* and *"C* as before, except that the common $2 shell costs are included, the sole change in the results is an increase in the intercept value *"a*, as

[16] For instance, see Jack L. Knetsch and Robert K. Davis, "Comparisons of Methods for Recreation Evaluation," in Allen V. Kneese and Stephen C. Smith, eds., *Water Research* (Baltimore: Johns Hopkins Press, for Resources for the Future, 1966), pp. 138–139.

reflected by line *JK*. And any specific number of visits per thousand, predicted by costs now, is also predicted by those same costs, less the $2 shell cost, in the earlier regression. For example, in figure 3, costs of $6, with $2 shell costs included, predict ten visits per thousand (on line *JK*), as do costs of $4 without the shell costs (on line *GH*). Visitation predictions are unchanged, as is the estimated economic value of a site; common costs are irrelevant.

The reasonableness of the cost categories used, for instance, in the travel cost approach, can be evaluated through criterion (1). If the only cost categories chosen for inclusion are transportation and the time cost of travel, there must be an implicit assumption that travel generates no satisfaction (or dissatisfaction) *per se*,[17] for only in this case may the cost differential be utilized to predict behavior of individuals when a fee generating no additional satisfaction is imposed. The validity of this assumption undoubtedly varies, depending upon the type of recreation under consideration. The assumption would seem invalid in the case of a family outing, and therefore upward bias would be imparted to the measured net benefits attributed to the destination site. On the other hand, the assumption may be close to the truth in the case of at least some hunters and fishermen whose choice of liquid refreshment causes them to travel in a state approaching anesthesia. The cost categories chosen carry an additional hidden assumption, namely, that all other expenditures exhibiting differentials between recreationists should be excluded from analysis. Again, the validity will differ according to the recreation. A fisherman en route may pay $15 for a motel room, yet may consider its niceties as giving him only $5 of direct satisfaction. Therefore $10 of costs from which no direct benefits are gained has been excluded improperly and a downward bias is imparted to measured net benefits.

The transfer cost approach, on the other hand, would include all of the $15 motel rental, inappropriately including the $5 of direct benefits, imparting an upward bias in the valuation of the resource or site in question. The transfer cost method assumes that an individual who prefers to fish from a rented yacht rather than a rented dinghy, while drinking cognac rather than applejack, and eating fresh beluga caviar rather than bologna, obtains no additional direct satisfaction from having incurred the higher costs. Only if this assumption is true (in general we believe it is false) can the cost differentials be utilized to predict behavior of others under an imposed fee from which no additional benefits are obtained.

[17] Should travel impart dissatisfaction, it would be treated in our terms as an increase in costs measurable in dollars. This allows criterion (1) greater generality.

The Pearse Method

A variant cost approach deserves mention. It was developed by Peter H. Pearse and was applied to a big-game hunting area in Canada.[18] In this approach, users of a recreational area are stratified on the basis of some factor or factors other than recreational costs (in Pearse's example, stratification was based on income). The person with the largest costs within a stratum is considered a marginal user who has received no net benefits from his recreational trip. All others within the same stratum are assumed to have obtained net benefits equal to the difference between their costs and those of the specified marginal user. The total net benefits determined depend to a large extent on the number of strata defined by the experimenter; net benefits will be zero if a stratum is defined for each recreationist, and will reach a maximum if all recreationists are considered as a single category. Deciding which of these measures coincides with the theoretical willingness-to-pay touchstone is problematical. Furthermore, if the diligent researcher stratifies both on the basis of differences in incomes and tastes, the possibility of downward bias of course increases as the number of strata increases.

Comparisons

Turning now to some similarities and contrasts between the Davis method and the cost techniques, particularly zonal travel–transfer cost, we find both require data obtained from the recreationist directly through interview or questionnaire, and both require information of a rather sensitive nature, e.g., the recreationist's income. It is quite often stated by supporters of the cost approach that hypothetical questions in the Davis technique yield hypothetical answers, while the alternative at least quizzes recreationists about their actual behavior—number of miles traveled, frequency and duration of visits, for instance.

The argument appears more substantive at a superficial level than at a deeper one. The statistically significant regression results obtained by Davis indicate that if the recreationists' answers were hypothetical, that term at least does not mean random or irrational. We shall have more to

[18] "A New Approach to the Evaluation of Non-Priced Recreational Resources," *Land Economics*, vol. 44 (February 1968), pp. 87–99. Pearse excluded on-site costs, and included imputed travel time costs, license fees, and "the declared cash cost of travel to and from the area" (p. 95). Whether the last mentioned is intended to include the costs of food and lodging en route, and whether hunters so interpreted whatever question they were asked, is not entirely clear. We have already made the point that license fees common to all play no useful role in determining the economic value of big game.

say about that matter in chapter 3. A researcher using the cost method either is going to have to make many assumptions, as yet untested, about how recreationists view the world, or he is going to have to ask the recreationist questions that surely will appear as strange as hypothetical ones. And if the researcher decides which cost categories should be included for analysis, he is making implicit assumptions regarding how recreationists view the benefits derived from the expenditures, as shown in the discussion of criterion (1). If an interviewer alternatively asks recreationists about division of direct and indirect benefits for a multitude of cost categories, he may observe as much bewilderment and resistance as he would if he asked a hypothetical question or two. Furthermore, given the possibility of biases entering the cost method, as discussed earlier, it is less than obvious that biases will be more severe if hypothetical questions are asked.

Given a choice of having a researcher assume appropriate values versus his obtaining such judgments from the recreationists themselves, we prefer the latter. Those who have conducted interviews on transportation costs per mile or travel time costs have often found the recreationists' perceptions at considerable variance with conventional assumptions. For instance, Stephen B. Mathews and Gardner M. Brown found that only 2 percent of sport salmon fishermen questioned, if precluded from fishing in a favored area, would work for pay instead.[19] Since a part of the fishermen's time presumably was spent traveling, the response results conflict with the conventional assumption that travel time should be valued in terms of wages forgone. The wage rate overestimates the opportunity cost of time and therefore produces an overestimate of the value of a site because working for pay ranked no higher than the third best alternative (third choices were not elicited). We personally believe that recreationists' own judgment of costs will serve to explain and predict better than will the imputation of cost magnitudes by the researcher.

All of the cost methods at their best can cope with willingness-to-pay net benefits, and hence generally will yield theoretically inappropriate underestimates of values where willingness to sell is the logical choice. For instance, the opportunity cost some may have borne in order to live close to a site, a value which should be measured where willingness to sell is appropriate, will not be captured by the cost methods. The Davis method does have the flexibility to cope additionally with willingness to sell. As an example, one might easily formulate a question asking individuals the minimum amount they need to be paid to give up for one year the right they now enjoy to recreate at Lake Success.

[19] *Economic Evaluation of the 1967 Sport Salmon Fisheries of Washington*, Technical Report No. 2 (Olympia: Washington Department of Fisheries, 1970), p. 13.

comparison of the three valuations for each individual as well as for the group as a whole. From individual to individual the value of V_D ranged below, between, and above the two values calculated from

$$\frac{V_{(D/Z)}}{Z} \text{ and } \frac{V_Z}{D/Z}$$

but the mean of V_D lay between the mean of the other two valuations.[7]

A difficulty arises with the use of these estimates: individuals reporting low magnitudes for D/Z, Z, or D are assigned relatively high marginal valuations, when in fact the low magnitude reported for any of the variables may indicate that the individuals were at or near unconstrained equilibrium, and in reality had a relatively low marginal net benefit from waterfowl.[8] Calculation of the marginal values by using the means of the untransformed independent variables circumvents the difficulty. These calculations are also shown in table 2.

Use of any of the valuations from equation (3-1) would pose a problem for work in chapters that follow. If more (or fewer) waterfowl were available and their value were to be approximated, a decision would have to be made regarding how much each margin was affected by the change. The value of V_D for equation (3-2), using the means of the independent variables, lies between the two values for equation (3-1) and does not pose the problem directly. Equation (3-2) is accepted as the prime valuation equation, and this equation's partial derivative V_D, evaluated at the means of the independent variables, is accepted as a first approximation

[7] A peculiarity arises from the multiplicative function form and the use of

$$\frac{V_{(D/Z)}}{Z} \text{ and } \frac{V_Z}{D/Z}$$

It is a simple exercise to show that if the two marginal values are taken for

$$V = p_0 Y^{p_1} S^{p_2} E^{p_3} (D/Z)^{p_4} D^{p_5}$$

then

$$\frac{V_{(D/Z)}/Z}{V_Z/(D/Z)} = \frac{p_4}{p_5}$$

Hence the two marginal valuations remain in constant proportion, not only for the mean valuations but for each individual calculation as well. Since p_4 and p_5 are assigned by the least-squares regression technique, the peculiarity presents a problem of little consequence when marginal values are taken for the sample group as a whole.

[8] A conflict with theory arises from the use of marginal valuations derived from an equation of the multiplicative functional form. The marginal valuation will always be positive, whereas in reality marginal valuations of zero would be encountered for individuals in unconstrained equilibrium. However, the difficulty does not seem great with the regression fittings used here. For instance, the valuations assigned to individuals under equation form (3-2) ranged from $32.21 to 33 cents. Hence the estimated marginal valuation could be quite low (as well as quite high).

of the marginal value of a waterfowl to a hunter, and hence as an approximation of a "market purchase price."[9]

Estimation with Costs Removed as an Independent Variable

Anthony Scott, among others, has objected to attempts to determine recreational values by questioning recreationists about the benefits they feel they have received:

> One [method] involves asking visitors what their benefits have been, or what they would demand to stay away. "Ask a hypothetical question and you get a hypothetical answer"; the results of this procedure have yielded extremely fanciful valuations. Worse, they show some respondents' inevitable misunderstanding of the question, for many valuations are close to their actually incurred expenses.[10]

Scott prefers valuation based on actual imposition of tolls on recreational sites. He does, however, admit that valuing migratory waterfowl hunting sites through "price and property markets" would be difficult.[11]

A question arises regarding what, precisely, is meant by a "hypothetical answer." As stated in chapter 2, a definition that such answers are random or irrational was not supported by the Davis experiment. Nor is it supported by ours, where respondents' estimates of consumer's surplus are significantly related to the independent variables at the 0.001 level.

Another definition of a "hypothetical answer" might be that responses are not related to the behavior that would be observed in the individual under market conditions. Obviously, this study cannot affirm or refute such a contention, and definitive resolution probably will occur only in an

[9] Were there a market for nontransferable rights to shoot and bag a waterfowl sold, let us say, by the United States government, the relevant price would be the equilibrium price, or the price that "cleared the market." Everyone who wished to buy *at that price* could do so, and precisely the number of rights to birds the government desired to sell would be sold. Each individual would purchase rights until the gross benefits from the last right purchased equaled the cost (sales price); marginal net benefits for each hunter would be zero. If at the last minute, after all rights had been sold and distributed, the government waived payment for the rights, each hunter would have a marginal net benefit from waterfowl in dollar terms equal to that of every other hunter.

The ideal valuation of a waterfowl from an economic standpoint would be the market-clearing price described above, but this cannot be determined accurately in the absence of a market. Under present institutional arrangements, marginal net benefits from waterfowl may vary from individual to individual. Thus a single value taken as the marginal net benefit can be only an approximation, and the value is not necessarily the market-clearing price.

[10] Anthony Scott, "The Valuation of Game Resources: Some Theoretical Aspects," *Canadian Fisheries Reports*, No. 4 (Ottawa: The Department of Fisheries of Canada, 1965), p. 37.

[11] *Ibid.*, p. 40.

experiment designed to match reported responses of recreationists against actual market behavior.

With Scott's objection regarding expenses and valuation in mind, equation (3-2) was regressed again, this time with costs omitted as an independent variable. The results were:

$$\ln V = 2.12 + \underset{(9.1)}{0.481 \ln Y} + \underset{(4.5)}{0.170 \ln S} + \underset{(13.5)}{0.428 \ln D} \qquad (t \text{ values}) \tag{3-6}$$

$$\bar{R}^2 = 0.199, \quad s^2 = 1.75, \quad n = 1511$$

The value of V_D for this equation, as shown in table 2, is moderately lower than the value when equation (3-2) is used. The correlation coefficient between costs and consumer's surplus in logarithmic transformation is a relatively low 0.22. On the basis of these results, the marginal valuation through the use of equation (3-2) is still accepted.

The Addition of 1967 Waterfowl Kill

Marion Clawson has hazarded the judgment that "more than half of the total satisfactions" of recreational experiences occur through recollection.[12] If so, it would appear that hunters may include in their estimates of net benefits from the present year's hunting the benefits they received this year from waterfowl bagged in previous years.

The hypothesis can be tested with data at hand, since the hunters in this study who were queried regarding their 1968 waterfowl kill were also queried a year earlier by the Bureau of Sports Fisheries and Wildlife as to their 1967 kill. The 1967 kill of each hunter who responded to the Bureau's survey was available in coded form. Each code number represented a range of kill, such as 6–10 birds. The means of the ranges were used, and the open-ended range, over 100 birds, arbitrarily was assigned the value of 125. A regression similar to that in equation (3-2) was run:

$$\ln V = 1.50 + \underset{(6.6)}{0.402 \ln Y} + \underset{(3.7)}{0.166 \ln S} \qquad (t \text{ values}) \tag{3-7}$$

$$+ \underset{(4.5)}{0.143 \ln E} + \underset{(7.2)}{0.315 \ln D} + \underset{(3.8)}{0.152 \ln D_{67}}$$

$$\bar{R}^2 = 0.228, \quad s^2 = 1.63, \quad n = 1122$$

The means of the variables used in equation (3-7) are reported in table 1. All variables are significant at the 0.001 level, which lends support to the

[12] Marion Clawson, "Economic Aspects of Sport Fishing," *Canadian Fisheries Reports*, No. 4 (Ottawa: The Department of Fisheries of Canada, 1965), p. 15.

spirit of Clawson's remarks. The reader may suspect that the closeness of the means for the waterfowl kill in 1967 and in 1968 indicates, that hunters report the same number of bagged waterfowl each year, regardless of actual kill. The correlation coefficient for the two variables in logarithmic form is 0.59, which does not seem high enough to support that hypothesis. Undoubtedly a true correlation does exist between the two variables.

Implications with respect to marginal valuation are interesting. The total marginal value of a waterfowl bagged in 1967, for instance, may be interpreted as the sum of its marginal value in 1967 and its discounted value in subsequent years. The partial derivative of V with respect to D_{67} in each year includes the regression parameter assigned to D_{67} as one of the multiplicative terms. From an examination of parameter values assigned to D and D_{67} in equation (3-7), it appears likely that the marginal value of D_{67} could be neglected after 2 or 3 years.

Presumably the D_{67} variable appears in the 1968 valuation function because hunters are conditioned by past success to appreciate hunting more than they would otherwise. Hence D_{67} may be interpreted as a taste variable reflecting a recreation value analog of learning by doing.[13] And by assuming tastes this year as given, regardless of their source, all values in this year should be attributed to this year's kill, contrary to our earlier interpretation. One may choose recollection value or learning-by-doing value, but not both, at least in this instance, without double counting.

COMPARISON OF QUESTIONNAIRE KILL DATA WITH THAT FROM OTHER SAMPLES

The only questionnaire results that may be compared to those of other studies are the responses regarding number of days hunted and kill. Table 3 shows data both from our questionnaire and from the U.S. Bureau of Sport Fisheries and Wildlife's annual waterfowl hunter surveys. Recall that our sample of 5,000 hunters, who answered our questions regarding 1968, was drawn from the list of 13,000 from which the Bureau derived its 1967 estimates for the seven western states.

Extrapolation shows the Bureau approximation of 1968 mean kill for successful hunters as 14.7 (row 1), while our sample reported 23.9 (row 4). The divergence requires investigation.

Are the individuals in our sample, and the respondents to our questionnaire, unrepresentative of the individuals who report to the Bureau? Apparently not, on the basis of kill figures for 1967. The Bureau's mean

[13] See Kenneth J. Arrow, "The Economic Implications of Learning by Doing," *Review of Economic Studies*, vol. 29 (July 1962), pp. 155–173.

Table 3. Mean Seasonal Retrieved Waterfowl Kill and Hunting Days for the Seven Western States

	1968	1967
(1) Mean waterfowl kill for the season for successful adult hunters in the seven states	14.7[a]	20.7[a]
(2) Mean waterfowl kill for the season for 2,720 successful adult hunters as reported to the Bureau by individuals among the 4,908 in this study's mailing list		19.9
(3) Mean waterfowl kill for the season for 1,366 successful adult hunters as reported to the Bureau by individuals among the 2,455 respondents to this study's questionnaire		19.9
(4) Mean waterfowl kill for the season for the 1,511 successful hunters (1968) in the regression sample, as reported in this study's questionnaire responses	23.9	
(5) Mean waterfowl kill for the season for the 1,122 successful hunters (1968 and 1967) in the regression sample		
(a) as reported in this study's questionnaire responses	25.0	
(b) as reported to the Bureau		25.7
(6) Mean number of hunting days for active adult hunters in the seven states[a]	6.9[a]	
(7) Mean number of hunting days for 1,691 active adult hunters, as reported in this study's questionnaire responses	9.7	

[a] Based on extrapolations from data in Charles H. Lobdell and Elwood M. Martin, "Waterfowl Harvest and Hunter Activity in the United States during the 1968 Hunting Season," Administrative Report No. 172 (Laurel, Md.: Migratory Bird Populations Station, U.S. Bureau of Sport Fisheries and Wildlife, 1969).

kill for that year (row 1) is consistent with that reported by individuals in our sample (rows 2 and 3), particularly considering that the data in the latter two rows were approximated using the Bureau's codes as outlined earlier.

The means in row 5, as would be expected, indicate that those respondents who were successful hunters in the two successive years either are superior hunters, hunt at superior sites, or hunt more days than the average hunter. Based on the 1967 information in rows 1 and 5b, let us assume that the 1,122 hunters actually did shoot on the average 5.0, or 24 percent, more birds than the average successful hunter in that year, and that those 1,122 hunters also shot 24 percent more birds in 1968.

Even if we assume that all 1,511 successful hunters in row 4, of which the 1,122 are a subset, shot 24 percent more birds in 1968 than the Bureau's mean kill of 14.7 in row 1, we still have 5.7 additional birds per hunter for which to account: $23.9 - (1.24 \times 14.7) = 5.7$.

These additional birds may be explained partially by sample bias. Since the names of those who purchased waterfowl hunting stamps in 1967 were used for this study of 1968 waterfowl hunting, new hunters in 1968 (presumably with lower mean kill) were not included, while discouraged 1967 hunters with low kill dropped out.

On first glance, it might appear that another explanation is available if the 1,511 individuals went hunting more days than average in 1968. Rows 6 and 7 show that active hunters in the study reported hunting 41 percent more days than the Bureau's mean. (Unfortunately, the Bureau's data cannot be extrapolated using successful hunters as the base.) But the row 3 figure implies that our respondents did not hunt more than the mean number of days for successful adult hunters in 1967, as the mean kill is not greater than that in row 1. Even if our hunters did hunt more days than average in 1968, the 24 percent adjustment already made on the basis of seasonal kill should have taken this into account.

Hunter surveys also are made by most of the seven western states. Regrettably, an examination of their results adds to rather than lessens the confusion. Almost without exception, the states' total kill estimates are greater than the federal government's. The states' estimates of average kill per hunter correspond more closely to the Bureau's than do those for total kill, however, indicating possibly that the divergence from federal results stems from the states' estimates of the number of waterfowl hunters.

The states inquire about waterfowl hunting as a part of a survey regarding all forms of hunting. Most of the states estimate the total number of waterfowl hunters (and indirectly the total waterfowl kill) through a simple expansion: if 10 percent of the questionnaire respondents say they hunted waterfowl, then it is concluded that 10 percent of all licensed hunters did so.[14] *If* waterfowl hunters are more conscientious in returning questionnaires, these hunters will be overrepresented in the response sample and an overestimate of the total number of waterfowl hunters will result. Additional problems occur because (1) the Bureau directly queries those who buy federal waterfowl hunting stamps, while the states query licensed hunters, a group which often includes substantial numbers of younger individuals who are not required to buy the federal stamp, and (2) a few of the states inquire about waterfowl "killed" rather than "killed and retrieved."

Adjustments might be attempted to circumvent all these difficulties, and to make the state estimates more comparable with the federal government's and ours. However, we suspect the contortions necessary to do so would be equalled only by the distortions in the product.

To summarize, after all plausible adjustments are made, average seasonal success in the federal survey seems to be 25 percent less than shown in our sample. It is useless to belabor bias problems further.

[14] We thank Walter F. Crissey for this point.

OTHER QUESTIONNAIRE RESULTS

The questionnaire elicited information on two matters not pertaining directly to waterfowl valuation, which were waterfowl preferences and reasons for not purchasing a waterfowl hunting stamp.

Waterfowl Preferences

Respondents who had purchased a 1968 waterfowl hunting stamp were asked to indicate their first choice in waterfowl as well as their second and third choices, if any. The results appear in table 4.

Table 4. Questionnaire Respondents' Waterfowl Preferences[a]

Response category[b]	First choice (percent)	Second choice (percent)	Third choice (percent)
Mallard	47.2	24.2	3.4
Geese	20.0	24.0	14.4
Ducks in general, no species preference	12.4	8.5	7.6
Pintail	10.0	20.3	12.5
American widgeon	0.9	3.7	7.4
Teal	0.8	4.9	9.2
Other varieties of ducks not listed above	0.5	0.6	1.1
Goldeneye, redhead, or scaup	0.1	1.2	1.5
Shoveler	—[c]	—[c]	0.5
Coots	0.0	0.0	0.6
No choice expressed	8.0	12.5	41.8

[a] The responses were counted only if the individual purchased a 1968 waterfowl hunting stamp; 2,013 potential waterfowl hunters are represented.

[b] The order of the response categories is not the same as that on the questionnaire.

[c] Less than 0.1 percent.

In some areas, at least, geese are a highly prized hunting trophy, presumably because of their beauty, size, and relative scarcity. Hence the preference shown for these birds was expected. The largest portion of the hunter bag on the Pacific flyway, however, consists of mallards and pintails, and strong preference was expressed for them. The preference is of no small import to this study, as those two species of dabbling ducks are prime summer occupants of the prairie pothole region.

Reasons for Not Purchasing a Waterfowl Hunting Stamp

Respondents who did not purchase a 1968 stamp were asked to check a reason for not doing so; the results are shown in table 5.[15] From an

[15] The categories used in the questionnaire were adapted from those in Thomas O. Kirkpatrick, *The Economic and Social Values of Hunting and Fishing in New Mexico* (Albuquerque: Bureau of Business Research, University of New Mexico, 1965), p. 84.

Table 5. Questionnaire Respondents' Reasons for Not Purchasing a 1968 Migratory Bird Hunting Stamp[a]

Response category[b]	Percent
Not enough time available to hunt	44.3
Insufficient number of waterfowl	11.5
Waterfowl hunting too expensive	7.0
Areas of waterfowl hunting too crowded with other people	6.6
Too far to drive to hunting area	4.1
Other reasons	14.3
Multiple reasons given	10.6
No reason given	1.6

[a] There were 442 respondents who did not purchase a stamp.
[b] The order of the response categories is not the same as that on the questionnaire.

economist's point of view, a person does not hunt when the benefits from waterfowl hunting are less than those of alternative uses of time and money. It is a gross simplification to ask a single reason for not hunting, when in fact many factors enter into the decision. Hence, the question was more an exercise in psychology than in economics.[16]

On that basis, the responses are of passing interest. The 1968 hunting season was considered a poor one, with lower than average duck production, and the seductive Canadian grain fields and mild fall temperatures caused many birds, especially mallards, to tarry in the northlands. Nevertheless, the plurality of respondents chose "not enough time" as the reason for not purchasing a stamp and not hunting, while only 11.5 percent chose "insufficient number of waterfowl."

[16] This question and the one regarding waterfowl preference appeared on the first page of the questionnaire. The possibility was not overlooked that these questions might appeal to the hunter and put him at ease.

chapter four Population Dynamics of North American Mallards

In the preceding chapters, attention was focused on waterfowl valuation, and, more indirectly, on demand for waterfowl. Here we turn to a complementary analysis of waterfowl supply. Since the processes involved are primarily those of nature, our framework switches from economics to biometrics.

Our discussion is couched not in terms of waterfowl or dabbling ducks, but in terms of mallards, the only species of waterfowl for which passable estimates of immature bird production are available. Fortunately these fowl provide an attractive example for this study, for a number of reasons. The biological requirements of several other species of dabbling ducks are similar to those of the mallard, suggesting that our approach may have broader applicability than available data allow us to demonstrate. Mallards are an important segment of waterfowl kill, and are major users of the wetlands of greatest concern, those in the prairies threatened by drainage. Furthermore, the availability of population data on a continental basis allows discussion of a tractable closed system.

THE THEORETICAL MODEL

Production Relationships

Population models for wildlife and fish state a relationship between the breeding population and the number of young that are produced, with

environmental elements added if they are critical. The importance of the quantity of wetlands available to support breeding mallards and their young was shown statistically by Walter F. Crissey.[1] The amount of wetlands is affected in the short run by the variability of climatological factors, including rainfall, and in the long run by man's creation and destruction of ponds.[2] Thus, the production function hypothesized for mallards is

$$I_t = f(W_t, P_t), \qquad I_{tW_t} > 0 \text{ and } I_{tW_tW_t} < 0 \tag{4-1}$$

where I_t is the number of immature mallards surviving to September in the given year t, W_t is the number of adult birds in May, and P_t is a measure of wetlands suitable for mallard production. To account for the signs of the partial derivatives, it is assumed over the relevant range that as the breeding population increases, the marginal or additional immatures produced become fewer and fewer for each equal additional increase in the number of breeders; and that the immatures exhibit compensatory mortality in the early weeks of their lives.[3]

Two alternative specific forms of equation (4-1) are considered. The first specific production function is the multiplicative relationship

$$I_t = AW_t^{\alpha} P_t^{\beta}, \qquad 0 < \alpha < 1, \quad 0 < \beta < 1 \tag{4-2}$$

where A, α, and β are constants. This form is recognized by economists as having general applicability in describing the relationship between the output of a firm or industry and the use of varying quantities of productive inputs. The equation implies that if equal successive amounts of either input, W or P, are added while the other input is held constant, each successive increase in output, I, will be less than the preceding one. Per-

[1] "Prairie Potholes for a Continental Viewpoint," *Saskatoon Wetlands Seminar*, Canadian Wildlife Service Report Series No. 6 (Ottawa: The Queen's Printer, 1969), pp. 161–171.

[2] The climatology affecting prairie wetlands is particularly complex. The region is semiarid. Melting snow provides the greater part of the water for the potholes. If the soil is wet and frozen during the spring runoff, the water does not soak readily into the soil and more of it reaches the depressions. Further moisture comes from thunderstorms during June and July. Long-term wet and dry cycles occur; one of the latter has been in evidence since the late 1950s. Sources: Allen G. Smith, Jerome H. Stoudt, and J. Bernard Gallop, "Prairie Ponds and Marshes," and John J. Lynch, "Weather," both in Joseph P. Linduska, ed., *Waterfowl Tomorrow* (Washington, D.C.: U.S. Government Printing Office, 1964), pp. 40, 42, 286–287.

[3] Compensatory mortality here means that the fraction of immatures that survive decreases as the number of immatures increases. The falling marginal function implies a falling average function, and a fall in the number of breeders is *compensated* by an increase in the average number of immatures per breeder. However, species survival contracts with nature are not unqualified. When the population level of some species falls below a critical minimum, the instantaneous death rate dominates the instantaneous birth rate and through depensatory mortality the species becomes extinct.

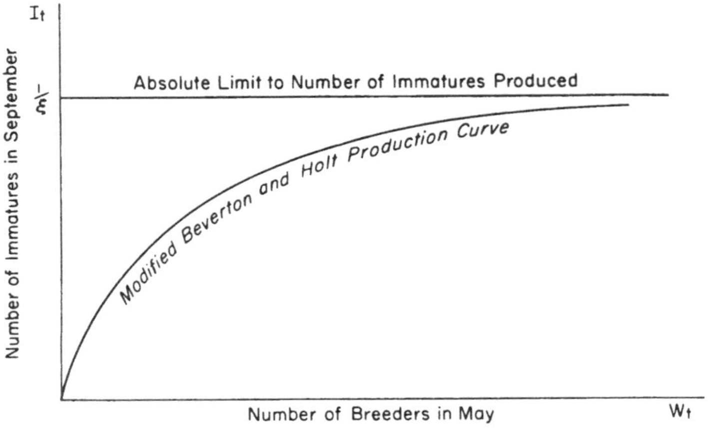

Figure 8. Modified Beverton and Holt equation for mallards illustrated for a single quantity of wetlands.

haps this economic "law of diminishing returns" also holds for some biological production relationships.

The second form is the Beverton and Holt relationship originally formulated for fish populations:

$$I_t = (k/W_t + \xi)^{-1}, \qquad \xi > 0, \quad 0 < k < 1 \qquad (4\text{-}3)$$

where k is a constant and ξ is defined below.[4] The Beverton and Holt relation can be modified for use with mallards through specification of ξ in two alternative ways:

$$\xi = (aP_t^b)^{-1}, \qquad a, b > 0, \text{ or} \qquad (4\text{-}4a)$$

$$\xi = ('a + 'bP_t)^{-1}, \qquad 'a, 'b > 0 \qquad (4\text{-}4b)$$

where a, b, $'a$, and $'b$ are constants. As W_t increases to infinity, I_t asymptotically approaches the limit of $1/\xi$, as can be seen from inspection of equation (4-3).

When equation (4-3) is used together with either equation (4-4a) or (4-4b), the asymptote will have the value aP_t^b or $'a + 'bP_t$ respectively. Hence each pond value defines a new curve relating W to I. Only one curve of this field is shown in figure 8.

Specification of (4-4b) results in equal increases in the asymptote with equal additional increases in the quantity of wetlands. Use of (4-4a) yields successively smaller increases in the asymptote under the same conditions.

[4] The equation is substantively the same as that in R. J. H. Beverton and S. J. Holt, *On the Dynamics of Exploited Fish Populations* (London: Her Majesty's Stationery Office, 1957), p. 49.

Equation (4-2), the multiplicative relationship, also defines a new curve relating W to I for each value of P. However, the multiplicative form of the equation indicates that for any given value of P, I will tend toward infinity as W tends toward infinity.

A Physical Balance Equation

As shown by the flow diagram in figure 9, a physical production function is only a part of our biological story. The remainder is told through an expression summarizing the survival of some of the birds from one year to the next:

$$W_{t+1} = S_2 (S_1 W_t + I_t - X_t) \tag{4-5}$$

Starting within the bracketed expression, only a fraction (S_1) of the May breeders (W_t) will survive to September. These birds, and their offspring (I_t), form the fall flight. A number (X_t) will be killed although not necessarily bagged by hunters. Of the remainder, represented by the entire bracketed expression, only a fraction (S_2) survive nature's vicissitudes to return to the breeding grounds the following spring. The present year is shown by t, the next by $t + 1$. During the hunting season it is assumed that mortality caused by hunters is so dominant that other sources of mortality may be neglected. It is further assumed that the instantaneous

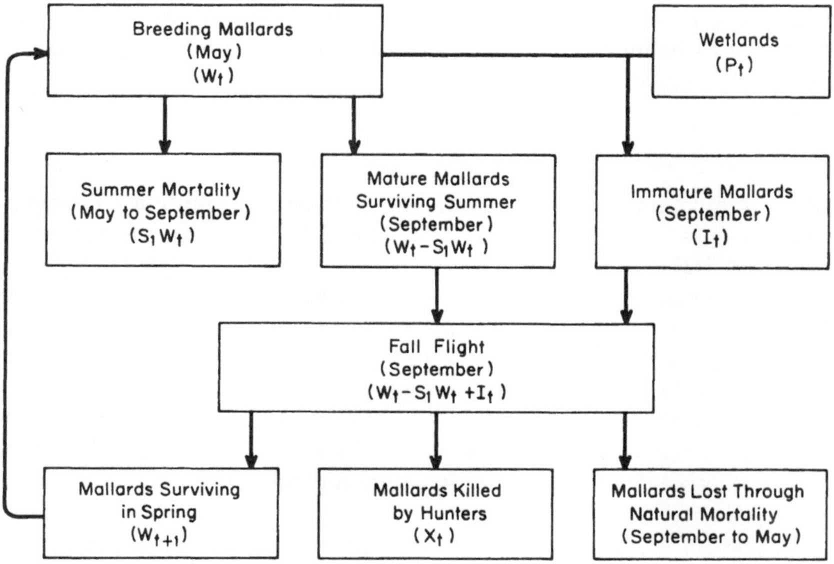

Figure 9. Flow diagram for the continental mallard population.

rate of subsequent natural mortality is independent of the instantaneous rate of hunting mortality. The implications of these mortality assumptions are discussed in a later section.

Steady-State Determination of Maximum Physical Yield

Either of the two production functions coupled with the dynamic or physical balance equation permit the identification of maximum physical sustained yield, under the restrictive condition that the quantity of wetlands is constant from year to year.

Let us demonstrate sustained yield through steady-state assumptions, using an asterisk to indicate a steady-state value. Hence

$$P_t = P_{t+1} = P^*, \quad W_t = W_{t+1} = W^* \tag{4-6}$$

and dropping time subscripts, equation (4-5) becomes

$$W^* = S_2 (S_1 W^* + I^* - X^*) \tag{4-7}$$

Rearranging terms,

$$X^* = I^* - \frac{1 - S_1 S_2}{S_2} W^* \tag{4-8}$$

Since the survival fractions, S_1 and S_2, are constants with values between zero and one, the fraction

$$\frac{1 - S_1 S_2}{S_2} > 0 \tag{4-9}$$

Define

$$c = \frac{1 - S_1 S_2}{S_2} \tag{4-10}$$

and substitute into equation (4-8),

$$X^* = I^* - cW^* \tag{4-11}$$

thus transforming the general physcial balance equation into specialized steady-state terms. It states that for a certain unvarying value of P, the number of mallards that may be killed each period without producing a change in the level of the May breeding stock is equal to the number of young birds in September minus the number required to compensate for adult attrition through natural death during the year.

Information from equation (4-11), combined with the modified Beverton and Holt production equation, is shown in figure 10. Since it is assumed that the quantity of wetlands is constant, there is only one Beverton and Holt curve, OV. Line OU, having the slope c, is derived from equation

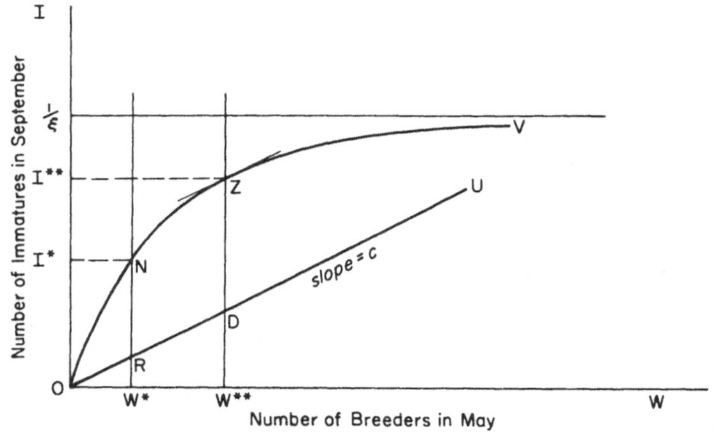

Figure 10. Modified Beverton and Holt production equation coupled with mortality information, determining steady-state mallard kill at various levels of breeding population.

(4-11). There is an infinite number of steady-state values for W^*, I^*, and X^*, one set being illustrated in the figure. Suppose the breeding stock is maintained at the level represented by OW^*. The Beverton and Holt curve determines that OI^* immature birds will be produced. To maintain the breeding population at the same level as the previous year, the number of birds killed must be equal to line length RN, which is equal to W^*N, the number of immatures produced, less W^*R, the number of birds dying of natural causes during the year.

Of the uncountable number of values of X^*, the largest will occur at that value of W^* where the slope of curve OV is equal to that of line OU. In figure 10, this occurs at Z, where the largest possible steady-state kill is equal to DZ, and the breeding population is OW^{**}, producing immatures equal to OI^{**}. Values associated with this maximum are designated by a double asterisk.

The value of the steady-state breeder population (W^{**}) which maximizes the steady-state kill also can be determined mathematically. Substituting the Beverton and Holt relationship (4-3) into equation (4-11) yields

$$X^* = (k/W^* + \xi)^{-1} - cW^* \tag{4-12}$$

The necessary condition for a maximum is

$$X^*_{W^*} = (k/W^* + \xi)^{-2}\,(k/W^{*2}) - c = 0 \tag{4-13}$$

Solving,

$$W^{**} = (-ck \pm \sqrt{ck})/c\xi \tag{4-14}$$

of which only

$$W^{**} = (-ck + \sqrt{ck})/c\xi \tag{4-15}$$

is positive and therefore relevant. The constants and the given number of ponds determine W^{**}. Substituting W^{**} into the Beverton and Holt equation (4-3) determines I^{**}, and then X^{**} can be found through equation (4-11).

The alternative, multiplicative production function (4-2) also may be substituted in equation (4-11):

$$X^{*} = AW^{*\alpha}P^{*\beta} - cW^{*} \tag{4-16}$$

The maximization proceeds in a fashion similar to that demonstrated previously,

$$X^{*}_{W^{*}} = \alpha AP^{*\beta}(W^{*})^{\alpha-1} - c = 0 \tag{4-17}$$

Solving,

$$W^{**} = (c/\alpha AP^{*\beta})^{\frac{1}{\alpha-1}} \tag{4-18}$$

and I^{**} and X^{**} can then be determined.

QUANTIFICATION OF THE MODEL

To apply the model represented by equation (4-11) and either of the production equations (4-2) or (4-3), data must be available for three variables and two parameters, namely, continental wetlands suitable to breeding mallards, number of adults in the spring, number of immature birds in the fall, the survival fraction of adults from May to September, and the survival fraction from September to May for mallards not shot by hunters.

Parameterizing the Production Relationships

Table 6 shows data relevant to mallards. The May continental breeding population estimates are derived from aerial breeding ground surveys. The number of immature birds in September is estimated through multiplying the breeding population figure by a production ratio calculated for each year.[5] At the time of our calculations data were limited to the 14 years shown. Aerial pond counts were made in central and southern Alberta, Saskatchewan, and Manitoba in May and again in July. Each of the two

[5] "In recent years this [production ratio] represents the age ratio in the kill adjusted for differential vulnerability to the gun as measured by the difference in recovery rate from preseason banding. Prior to 1960 the ratios were derived in part from bag check records and in part by calculations." From Crissey, "Prairie Potholes," p. 163.

Table 6. Mallard Population Statistics and Canadian Prairie Pond Indices (millions)

Year	Continental mallard breeding population, May	Number of immature birds, September	Ponds, May	Ponds, July
1955	11.0	15.6	6.0	5.0
1956	12.2	13.9	4.6	3.6
1957	11.6	18.7	2.8	2.2
1958	14.5	11.0	3.1	1.7
1959	11.6	6.6	1.9	1.2
1960	9.7	11.0	3.7	1.7
1961	9.0	6.0	1.7	0.6
1962	6.9	6.6	2.3	0.8
1963	8.0	7.6	2.4	1.7
1964	7.9	6.0	2.6	1.1
1965	6.1	9.3	3.5	2.2
1966	7.8	8.9	3.6	2.0
1967	8.6	9.8	3.7	1.5
1968	8.3	6.2	1.7	0.85

Source: Estimates from Migratory Bird Populations Station, U.S. Bureau of Sport Fisheries and Wildlife, Laurel, Maryland. All data except for the year 1968 are published in Aelred D. Geis, R. Kahler Martinson, and David R. Anderson, "Establishing Hunting Regulations and Allowable Harvest of Mallards in the United States," *Journal of Wildlife Management*, vol. 33 (October 1969), pp. 848–859.

counts is a possible substitute measure for the unknown quantity of wetlands relevant to mallard production.

It may be hypothesized that the July pond count is superior, for it is small comfort to all but the most existential ducklings hatched "on" May ponds if the pond disappears soon thereafter. Ducklings require a high-protein diet derived primarily from aquatic invertebrates,[6] and a brood on a pond drying up in early summer must be moved, with resulting high mortality. There is a hypothesis (or perhaps folklore) having considerable currency that mallards often are able to sense and to avoid ponds that will dry up during the summer. If this is ever proved, it will be an intriguing justification for the use of July pond counts. Mallards determine in the spring which wetlands will remain in July, and act accordingly.

The multiplicative relationship (4-2) lends itself to linear regression after the variables have been converted to logarithms, and it is worthwhile to compare the statistical results using alternatively the May and the July pond data. Using the statistics in table 6, the following results were obtained:

$$\ln I = 1.07 + 0.413 \ln W + 0.479 \ln J \qquad (4\text{-}19)$$
$$(t \text{ values}) \qquad (1.7) \qquad (4.7)$$
$$\bar{R}^2 = 0.72, \; s^2 = 0.039, \quad n = 14$$

[6] Linduska, *Waterfowl Tomorrow*, p. 45.

where J represents the July pond data, and

$$\ln I = 0.159 + 0.600 \ln W + 0.672 \ln Q \qquad (4\text{-}20)$$
$$(t \text{ values}) \qquad (2.4) \qquad (4.1)$$

$$\bar{R}^2 = 0.66, \quad s^2 = 0.047, \quad n = 14$$

where Q represents the May pond data. In both regression fittings an earlier test by W. J. Dixon on the I residuals requires rejection of the data for the year 1957 at the 0.10 level.[7]

The regressions were run again, with 1957 data omitted. The results using July ponds were:

$$\ln I = 1.36 + 0.269 \ln W + 0.460 \ln J \qquad (4\text{-}21)$$
$$(t \text{ values}) \qquad (1.6) \qquad (6.7)$$

$$\bar{R}^2 = 0.83, \quad s^2 = 0.018, \quad n = 13$$

with no further outliers. Using May ponds brought forth another outlier, and with the 1964 data also omitted, the final results were:

$$\ln I = 0.7 + 0.7 \ln W + 0.3 \ln Q \qquad (4\text{-}22)$$
$$(t \text{ values}) \quad (4.1) \qquad (6.6)$$

$$\bar{R}^2 = 0.98, \quad s^2 = 0.003, \quad n = 12$$

For the biological reasons stated earlier, July ponds are accepted as the wetland measure for the remaining biometric work. However, the coefficients in May pond regression (4-22) exhibit constant returns to scale, a property so seductive to economists that the results will be used alternatively in a following chapter.

The inability of the July pond regression to cope with 1957 data (when immature production was exceedingly high) probably indicates that a relevant independent variable has been left out of the regression. Since waterfowl biologists have speculated that spring weather is an important factor in explaining dabbling duck production, an attempt was made to incorporate mean minimum temperatures, mean daily temperatures, number of days of freezing temperatures, and number of days of rain, for six Canadian prairie weather stations, into equation (4-19), which includes the 1957 production data. The regression coefficients associated with the weather varieties were not statistically significant, and the missing variable problem remains unsolved.

The modifications of the Beverton and Holt formulation represented by equations (4-3), (4-4a), and (4-3), (4-4b) were fitted using a least-squares nonlinear regression format. When data for all 14 years were used, the 1957 data were rejected by the outlier test at the 0.10 level.

[7] "Processing Data for Outliers," *Biometrics*, vol. 9 (March 1953), pp. 74–89.

With the 1957 data omitted, the regression results for equation forms (4-3) and (4-4a) combined were

$$I = [0.297/W + (9.28J^{0.660})^{-1}]^{-1} \quad (4\text{-}23)$$

$$\bar{R}^2 = 0.88, \quad n = 13$$

and the results for equation forms (4-3) and (4-4b) combined were

$$I = [0.361/W + (4.16 + 5.82J)^{-1}]^{-1} \quad (4\text{-}24)$$

$$\bar{R}^2 = 0.87, \quad n = 13$$

There were no further outliers in either case.

Neither on *a priori* grounds nor on the basis of regression results is it possible to choose a single mallard production equation from the multiplicative relationship or the two Beverton and Holt modifications. Therefore all three remain under consideration in this chapter.

Values for Constants in the Steady-State Equation

To animate steady-state balance equation (4-11), two values are required: the fraction of adult mallards surviving during the summer months (S_1), and the fraction of mallards not shot by hunters surviving during the rest of the year (S_2).

The value of S_1 is estimated at 0.95 to 0.96;[8] the magnitude 0.95 is used throughout. It is more difficult to settle on a value for S_2. Natural mortality figures for mallards during the fall, winter, and spring period combined have been estimated by the Migratory Bird Populations Station, U.S. Bureau of Sport Fisheries and Wildlife. The natural mortality is calculated by subtracting from the estimated fall mallard population both the estimated annual Canadian and United States kill and the breeding population of the subsequent spring; hence, natural mortality is not estimated independently. When the mean number of birds thought to have died from natural causes during the fall, winter, and spring for the years 1955–65 is divided by the mean number of birds estimated to have been in the fall population over the same years, the natural mortality fraction is 0.16, which would give a crude estimate of S_2 at 0.84. The natural mortality fraction estimated for each of the years ranges from 0.07 to 0.27.[9]

In view of the imprecision of the estimates, two alternative values of S_2 are used in the computations that follow, namely, 0.80 and 0.85. This determines values of c in equation (4-11) of 0.300 and 0.266 respectively. Although S_2 may be neither of the figures used, they demonstrate the

[8] Crissey, "Prairie Potholes," p. 163.
[9] *Ibid.*, p. 163.

sensitivity of the model to changes in the magnitude of this survival fraction.

Steady-State Values

There are two forms of population production equations which are multiplicative and two Beverton and Holt variations to be combined with steady-state equation (4-11), for which two estimates of S_2 are proposed. Hence there are six combinations. Sets of computations appear in tables 7 and 8 for the July pond value of 1.4 million, the approximate mean number of July ponds in the relatively dry period 1958–68 inclusive. Table 7 shows values for W^{**}, I^{**}, and X^{**} which are considered estimates of biological optimums in the sense of maximizing the number of birds that may be killed, given the population equation, the mean July pond count, and the value for S_2. A broader range of steady-state values is shown in table 8 for immatures and kill resulting from selected numbers of breeders.

Table 9 is similar in format to table 8 except that a wet period is simulated through assuming a July pond count of 3.0 million. To dam the inundation by digits, only the S_2 value of 0.80 is used.

Comparison of results in both tables shows that at relatively low levels of breeding mallards the multiplicative production equation estimates higher immature production, and hence also allows higher annual harvest than do its Beverton and Holt counterparts. These relationships are reversed at relatively high levels of the breeding population. However, differences are minor, and data for many more years will be needed to demonstrate convincingly the superiority of one of the production relationships.

In table 8, changing the natural survival fraction S_2 from 0.80 to 0.85 allows moderate increases in the annual kill for any given level of breeders. Furthermore, the number of breeding adults needed to obtain maximum physical sustained yield increases.

Table 7. Steady-State Maximum Sustained Yield Solutions Assuming July Ponds Are a Constant 1.4 Million (millions)

	Multiplicative production relationship		Beverton and Holt using aP^b		Beverton and Holt using $'a + 'bP$	
	$S_2 = 0.80$	$S_2 = 0.85$	$S_2 = 0.80$	$S_2 = 0.85$	$S_2 = 0.80$	$S_2 = 0.85$
W^{**}	6.86	10.11	8.09	9.83	9.06	11.10
I^{**}	7.66	8.49	8.13	8.58	8.26	8.79
X^{**}	5.60	6.21	5.70	6.36	5.54	6.28

Table 8. Steady-State Values for Immatures and Kill at Various Magnitudes of Breeding Mallards, Assuming July Ponds Are a Constant 1.4 Million (millions)

	Multiplicative production relationship			Beverton and Holt using aP^b			Beverton and Holt using $'a + 'bP$		
Breeders	Immatures	Kill $S_2 = 0.80$	Kill $S_2 = 0.85$	Immatures	Kill $S_2 = 0.80$	Kill $S_2 = 0.85$	Immatures	Kill $S_2 = 0.80$	Kill $S_2 = 0.85$
5	7.03	5.53	5.90	6.87	5.37	5.73	6.51	5.01	5.38
6	7.39	5.59	6.03	7.37	5.57	6.01	7.07	5.27	5.71
7	7.70	5.60	6.11	7.77	5.67	6.18	7.52	5.42	5.94
8	7.98	5.58	6.17	8.10	5.70	6.29	7.91	5.51	6.10
9	8.24	5.54	6.20	8.38	5.68	6.35	8.23	5.53	6.20
10	8.47	5.47	6.21	8.62	5.62	6.36	8.52	5.52	6.25
11	8.69	5.39	6.20	8.83	5.53	6.34	8.76	5.46	6.27
12	8.90	5.30	6.18	9.01	5.41	6.29	8.98	5.38	6.26

Table 9. Steady-State Values for Immatures and Kill at Various Magnitudes of Breeding Mallards, Assuming July Ponds Are a Constant 3.0 Million and Survival Fraction S_2 Is 0.80 (millions)

	Multiplicative production relationship		Beverton and Holt using aP^b		Beverton and Holt using $'a + 'bP$	
Breeders	Immatures	Kill	Immatures	Kill	Immatures	Kill
8	11.3	8.93	11.2	8.80	10.9	8.54
9	11.7	8.99	11.7	9.04	11.6	8.87
10	12.0	9.03	12.2	9.22	12.1	9.13
11	12.3	9.04	12.6	9.33	12.6	9.34
12	12.6	9.03	13.0	9.40	13.1	9.49
13	12.9	9.01	13.3	9.43	13.5	9.60
14	13.2	8.97	13.6	9.43	13.9	9.67
15	13.4	8.91	13.9	9.40	14.2	9.71
16	13.6	8.85	14.1	9.34	14.5	9.72
17	13.9	8.77	14.4	9.26	14.8	9.71

Perhaps the most interesting result in the tables is the insensitivity of kill magnitude to changes in the number of breeders, holding the pond count constant. For instance, in table 8, using the Beverton and Holt function in columns four and five, and the S_2 value of 0.85, increasing breeders 100 percent from 5 million to 10 million yields a harvest increase of only about 11 percent. On the other hand, increasing the number of ponds while holding the number of breeding adults constant had a considerably greater effect on kill. For example, going from table 8 with 11 million breeders, the same Beverton and Holt equation, and the S_2 value of 0.80, to their counterparts in table 9, we find that a 114 percent increase in ponds yields a 69 percent increase in harvest.

Release from Steady State Through Simulation

The identification of maximum sustained yield requires the assumption that the number of wetlands remains constant from year to year. A quick glance at table 6 demonstrates the lack of reality of this assumption. In fact, the variability in pond counts seems to fit the components of classic time series analysis: long-run trend, and cyclical, seasonal, and irregular fluctuations.

Whatever the trend, which is undoubtedly influenced by long-run climatological changes and by geomorphic alterations through natural and human agencies, the possibilities of quantification from 14 years of Canadian pond data are slim indeed, and hence trend is neglected. Seasonal fluctuation is eliminated by the expedient of using data for the same season each year, i.e., July ponds. Prairie ponds are believed to show cyclical

influences, and the data in table 6 may be interpreted as illustrating a swing from a wet cycle to a dry one, with irregular components also contributing to the observed figures.

The effects of irregular (and theoretically, cyclical) components can be studied through computer simulation. In such fashion, mallard harvest where the number of ponds varies from year to year can be compared to the harvest under steady-state conditions. Actual pond data, for the 14 years available, contain relatively little information about a wet cycle, and therefore our reported results will be confined to dry-cycle simulation where irregular fluctuations, assumed to be random and normally distributed, are allowed full play. Since immature production and ponds are undoubtedly subject to irregular fluctuation, the Beverton and Holt production functions chosen for stochastic simulation are:

$$I_t = \gamma_2\{0.297/W_t + [9.28(\gamma_1 J)^{0.660}]^{-1}\}^{-1}, \qquad 0 < \gamma_1, \gamma_2 < 2 \qquad (4\text{-}25)$$

and

$$I_t = \gamma_4\{0.361/W_t + [4.16 + 5.82(\gamma_3 J)]^{-1}\}^{-1}, \qquad 0 < \gamma_3, \gamma_4 < 2 \qquad (4\text{-}26)$$

which obviously utilize the earlier regression estimates for both the aP^b and $'a + 'bP$ variations. The several γ's are normally distributed random variables with means of one, and standard deviations yielding pond and immature values consistent with the dry-cycle years of table 6. These standard deviations are approximated through trial and error at 0.330 for both γ_1 and γ_3, and 0.033 for γ_2 and γ_4. The value given J is 1.4 (million ponds), the same as used in tables 7 and 8.

Coupling either (4-25) or (4-26) with the physical balance equation (4-11) allows simulation of two selected management schemes. In the first, the fraction of the fall flight that may be killed under maximum physical sustained yield conditions is determined. Then, this fraction of the fall flight is assumed killed in each year even though the simulated fall flight varies in size from year to year as a result of the random influences. In the second plan, absolute kill magnitude is set so that the number of breeders in the following spring will be equal to W^{**}. (With random influences, it is at least theoretically possible that immature production will be so low that no kill whatsoever is allowed. In such case, and only then, the model allows next spring's breeders to be whatever number naturally survive, a magnitude that may be less than W^{**}.)

Both plans were run for five hundred 30-year chains, each chain starting with the optimal breeding population, W^{**}. The results, using the value of 0.84 for S_2, are summarized for convenience in an informal table:

Beverton and Holt production function	Management plan 1		Management plan 2	
Using form aP^b	220,000	3.5%	191,000	3.1%
Using form $'a + 'bP$	122,000	2.0%	106,000	1.7%

where the absolute values represent the amount the average annual harvest, in stochastic simulation, falls below the steady-state maximum physical sustained yield, and the remaining values show the absolute values as a percentage of the maximum sustained yield.

It may be argued that the 30-year chains are too long, for wetland dry cycles historically do not last that long. Simulation shows that reduction in the length of the chain causes reduction of all magnitudes in the table. As stated earlier, lack of real-world data precludes our having much confidence in the realism of simulation runs made for wet cycles, or for wet and dry cycles combined, but none of our results from attempts at such runs conflicts with the spirit of the numerical results reported above.

Throughout our experiments, including those reported in the table, the second management plan shows what we can only characterize as consistent trivial superiority. Far more importantly, we conclude that under either management plan the introduction of random influences in ponds and immature production has a very modest effect on average kill when compared to optimal steady-state magnitudes.

COMPARISON WITH A DIFFERENT MODEL

Summary of Another Approach

Our model can be compared to that used by the U.S. Bureau of Sport Fisheries and Wildlife to determine allowable mallard harvest and waterfowl hunting regulations. In their model, as expounded by Geis, Martinson, and Anderson, the number of mallard young in the fall is predicted through a simple linear regression equation where independent variables account for information regarding July and May pond data, the breeding population, and an index to the number of broods in July.[10] The resulting immature bird estimate is combined with the estimated number of breeders surviving the summer in order to determine the size of the fall flight. Subjective decision gives the desired breeding population for the coming

[10] Aelred D. Geis, R. Kahler Martinson, and David R. Anderson, "Establishing Hunting Regulations and Allowable Harvest of Mallards in the United States," *Journal of Wildlife Management*, vol. 33 (October 1969), pp. 848–859.

spring, and then the total number of mallards that may die from all causes between this fall and the next is obtained.

For Geis *et al.* to move from this last statistic to the appropriate size of the mallard harvest requires use of a functional relationship between the fraction of the fall flight lost through hunter mortality and the fraction of the fall flight dying from all causes by May of the following year. This functional relationship, attributed to Hickey,[11] is assumed by Geis *et al.* specifically to have linear form, or,

$$\frac{T_\tau}{F_\tau} = a' + b'\left(\frac{X_\tau}{F_\tau}\right) \tag{4-27}$$

where X_τ is the number of mallards killed but not necessarily bagged by hunters, F_τ the fall flight, and T_τ the total number of mallards dying from all causes during a year. The biological year is denoted by a different symbol, τ, than has been used previously. The Hickey–Geis *et al.* equation starts the year on September 1, while we have implicitly started the year at the beginning of May. The constants a' and b' are estimated at 0.334 and 0.59 respectively by Geis *et al.* through linear regression and other data manipulation.

As previously noted, Geis *et al.* determine figures for T_τ and F_τ. By working backward through equation (4-27) they obtain the fraction of the fall flight that may be harvested and from that is derived the actual harvest number. Hunting regulations are then set to approximate the desired kill.

Recasting for Comparison

The Geis *et al.* method has more in common with our model than is obvious at first glance. In fact, a simple exercise shows our model to be entirely consistent with Hickey–Geis's linear equation (4-27). This recasting allows us to demonstrate some rather well-hidden assumptions and relationships underlying both models.

The information in equation (4-5), modified to start the year in September rather than May, provides our starting point:

$$W_\tau = S_2(S_3 W_{\tau-1} + I_\tau - X_\tau) \tag{4-28}$$

The year, τ, is divided into three subperiods: Δ_1, from September 1 to the end of the hunting season; Δ_2, from the end of the hunting period to May 1; and Δ_3, from May 1 to September 1. In terms of the flow diagram (figure 9), our year starts with the fall flight, and all other relationships on the diagram follow sequentially. For simplicity it is assumed that the

[11] Joseph J. Hickey, *Survival Studies of Banded Birds*, Special Scientific Report Wildlife No. 15 (Washington, D.C.: U.S. Bureau of Sport Fisheries and Wildlife, 1952).

waterfowl hunting season on the North American continent starts at the beginning of subperiod Δ_1, and that during this season, mortality caused by hunters is so dominant that other sources of mortality may be neglected.

In the equation above, the summer survival fraction S_3 is exactly the same as that denoted earlier by S_1; the subscript has been changed so that it coincides with the time period to which it is related. The survival fraction S_2 also coincides (without modification) to its time period, as do the mortality fractions that we now denote as M_2 and M_3. Obviously the natural survival and mortality fractions for either subperiod must add up to one, and the mortality fraction can be expressed as one minus the survival fraction.

Now, in our model,

$$F_\tau = S_3 W_{\tau-1} + I_\tau \tag{4-29}$$

that is, the fall flight is composed of the adults surviving the summer plus the September immatures. From this definition and the information in equation (4-28),

$$T_\tau = X_\tau + M_2(F_\tau - X_\tau) + M_3 S_2(F_\tau - X_\tau) \tag{4-30}$$

that is, the total number of birds dying in the annual period is equal to the hunter harvest, plus the number of the fall flight remaining after the harvest that die from natural causes by the following May, plus the number of May breeders, represented by $S_2(F_\tau - X_\tau)$, which die during the summer.

Elementary rearrangement of (4-30) yields

$$\frac{T_\tau}{F_\tau} = (M_2 + M_3 S_2) + (1 - M_2 - M_3 S_2)\frac{X_\tau}{F_\tau} \tag{4-31}$$

Noting that

$$S_2 = (1 - M_2) \tag{4-32}$$

then (4-31) may be restated entirely in terms of mortality fractions, or

$$\frac{T_\tau}{F_\tau} = (M_2 + M_3 - M_2 M_3) + \left[1 - (M_2 + M_3 - M_2 M_3)\frac{X_\tau}{F_\tau}\right] \tag{4-33}$$

The restatement of our model in terms of the Hickey–Geis relationship is now complete. Since we have assumed that the natural mortality fractions are constant, the intercept value expressed by the first bracketed term in (4-33) is the equivalent of Hickey–Geis a', while the expression in square brackets is equivalent to their slope value b'. Obviously in (4-33) the slope value is equal to one minus the value of the intercept. Note that if hunter kill is zero, the fraction of the fall flight dying is determined

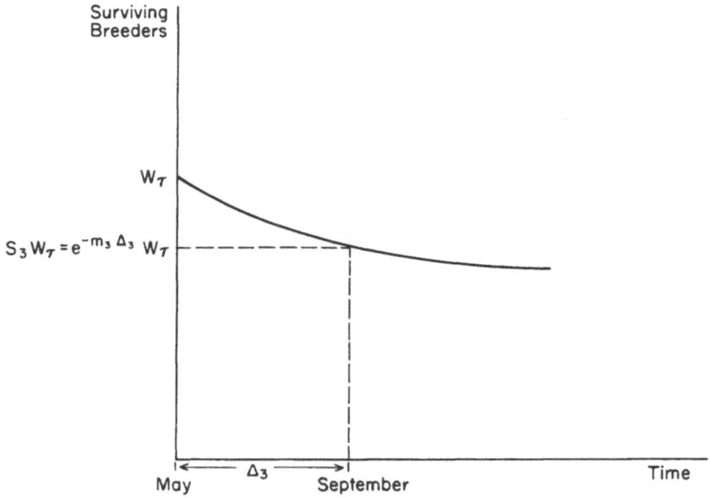

Figure 11. Graphical representation of effects of instantaneous mortality rate over time.

solely by constant natural mortality fractions. There is, however, an internal difficulty posed by the assumption that natural mortality fractions are constant. To bring this problem to light requires the expression of survival and mortality fractions in terms of instantaneous rates.

The fraction of the May breeding population surviving to September, for instance, can be expressed as

$$S_3 = e^{-m_3\Delta_3} \tag{4-34}$$

where m_3 is the instantaneous mortality rate. Figure 11 shows the initial May breeding population as W_τ on the vertical axis. Throughout the subperiod Δ_3, the birds die at the continuous instantaneous rate expressed through m_3, leaving fewer birds surviving at the end of the subperiod as shown. Obviously,

$$M_3 = 1 - e^{-m_3\Delta_3} \tag{4-35}$$

the fraction of birds dying during the subperiod is equal to one minus the fraction surviving. A corresponding relationship holds between M_2 and m_2.[12]

[12] Additional treatment of survival and mortality expressions is given in W. E. Ricker, *Handbook of Computations for Biological Statistics of Fish Populations* (Ottawa: The Queen's Printer, 1958).

So that all the instantaneous rates are explicit, we note parenthetically that

$$X_\tau = (1 - e^{-h\Delta_1})F_\tau \tag{4-36}$$

where h is the instantaneous rate of hunting mortality, undoubtedly influenced by bag and possession limits and by the number of hunters. Independence of the instantaneous hunting and natural mortality rates is assumed. This, for example, implies that changing the bag limit will not affect instantaneous natural mortality rates later coming into play.

With these preliminaries out of the way, let us return to our contention of inherent difficulties in the assumption of constant natural mortality fractions. From earlier discussion,

$$M_2 = 1 - e^{-m_2\Delta_2} \tag{4-37}$$

But Δ_2, the subperiod immediately following the hunting season, varies inversely with the length of that season. Hence assuming M_2 constant implies rather heroically that any change in Δ_2 is exactly offset by a change in m_2 so that $e^{-m_2\Delta_2}$ has constant value. For our work, the problem is of little consequence. Both in this chapter and in those following our emphasis is on steady-state solutions, where the assumption of a constant hunting season length, and therefore a constant Δ_2, is felicitous.

Geis *et al.* make no explicit assumption regarding instantaneous mortality rates in their use of equation (4-27). Since we know of no set of assumptions more consistent with reality that produce a singular linear relationship for their equation, it seems reasonable to conclude that they must somehow explain why increasing the length of the hunting season increases the instantaneous rate of natural mortality during the subsequent period. And they face an additional problem: their estimate values for a' and b' do not sum to the theoretically appropriate 1.0, but rather to 0.924.

Altering Mortality Assumptions

Additional difficulties come to light with respect to the Hickey–Geis relationship when the assumption is made that natural mortality takes place during the entire year, including the hunting season, at a constant instantaneous rate, m. In this case the only fractional period of a year needing explicit notation is the hunting season Δ_1, and

$$X_\tau = \frac{h}{h+m}[1 - e^{-(h+m)\Delta_1}]F_\tau \tag{4-38}$$

The bracketed expression is the fraction of the fall flight that will die from all causes during the hunting season. The additive relationship

between h and m implies independence of the instantaneous rates. In any instant, the mortality of mallards from hunting and the mortality of mallards from all causes is in the ratio $h/h + m$. Since both h and m operate over the entire subperiod under consideration, the ratio also represents the fraction of total deaths due to hunting during Δ_1. Also,

$$\frac{X_\tau}{F_\tau} = \frac{h}{h+m}[1 - e^{-(h+m)\Delta_1}] \tag{4-39}$$

It is true that

$$\frac{T_\tau}{F_\tau} = 1 - e^{-(h\Delta_1 + m_\tau)} \tag{4-40}$$

Omitting the τ in the exponent of (4-40) since it is equal to one, and substituting (4-39) and (4-40) into the Hickey–Geis relationship (4-27) yields

$$1 - e^{-(h\Delta_1 + m)} = a' + b'\left\{\frac{h}{h+m}[1 - e^{-(h+m)\Delta_1}]\right\} \tag{4-41}$$

Assume that the relationship holds in the absence of hunting. Then

$$a' = 1 - e^{-m} \tag{4-42}$$

and the vertical intercept is of course the natural mortality fraction. Substituting (4-42) into (4-41) and solving for b',

$$b' = \frac{e^{-m} - e^{-(h\Delta_1 + m)}}{\dfrac{h}{h+m}[1 - e^{-(h+m)\Delta_1}]} \tag{4-43}$$

The slope coefficient is explicitly related to the instantaneous rate of hunting mortality and to the length of the hunting season. Now b' must be constant in order to obtain a single linear relationship between the fraction of the fall flight dying from all causes and the fraction of the fall flight killed by hunters. And for that to be true, one of two assumptions must be made: either (1) h and Δ_1 are assumed to be constants, or (2) h and Δ_1 are assumed to be affected in a most complex fashion, whereby a change in h, for example, is precisely offset by the change in Δ_1, so the value of b' remains unchanged. Since bag and possession limits, both affecting h, are changed from year to year, as is the hunting season length, assumption (1) is incorrect and assumption (2) strains credulity.

Since neither of the assumptions is realistic, changing hunting regulations therefore changes the slope of the line, and the Hickey–Geis simple linear relationship does not hold in this more sophisticated setting. Data points for years having differing hunting regulations, if plotted on a graph showing the fractions of the fall flight dying from all causes and the frac-

tion of the fall flight killed by hunters, will all lie on *different* lines. Forcing single regression line through such points is an exercise in futility.

Conclusion

Our emphasis is upon explanation, while Geis *et al.* are primarily concerned with prediction. This leads to some difference in approach. For instance, their production equation for immature fall mallards includes a summer brood count as one of the independent variables. This may be an excellent choice for their purposes, but it is totally inappropriate for ours, as we then would have to explain the magnitude of the brood count. Additional differences arise because our time horizons are longer than those of individuals concerned with the annual setting of bag and season limits. Fortunately for us, our interest in, and our ability to use, steady-state solutions allows us to bypass difficulties discussed earlier with respect to the mortality assumptions. Those concerned with the determination of hunting regulations are less favorably situated to evade the dilemmas posed.

chapter five Optimal Management Programs

We begin with a crude cost–benefit analysis of existing wetlands. Although analytically nonoptimal, its simplicity has merit and the results are accurate up to an order of magnitude—or so we feel. Often this level of accuracy is satisfactory. Some of the biometric information of chapter 4 will be joined with an earlier marginal valuation figure for waterfowl to indicate whether preservation of wetlands threatened by drainage is economically justifiable. Our exercise will use information from government wetlands programs.

WETLANDS PURCHASE AND EASEMENT PROGRAMS

Both Canada and the United States have government-supported programs for wetlands purchase and easement. Although some outright purchases of land have been made under the Canadian program, its major thrust has been the acquisition of easements requiring that the wetlands be left undisturbed for the use of waterfowl. The easements are temporary, generally running for a 10-year term. An annual payment is made, representing 6 percent of the appraised value of the surrounding land. At first all wetland types were accepted in the program, but more recently they have been restricted to permanent types. Average annual payments made on recently acquired easement land are \$4.73 per acre for the Canadian

prairie as a whole, and \$5.60 per acre for land in Alberta.[1] The latter area produces many waterfowl shot by Pacific flyway hunters in the United States.

The U.S. program is financed by funds from sales, and future expected sales, of waterfowl hunting stamps. Breeding-ground purchases have been made predominantly in North Dakota, South Dakota, and western Minnesota. Only permanent wetlands are considered for the purchase portion of the program. Recent wetland values in the area range from \$25 per acre to \$85 or so. Studies indicate an average production of 2.7 birds per acre of wetlands in the program.[2]

COST–BENEFIT COMPUTATIONS

Let us be incautious and assume that \$3.29 marginal value of a bagged waterfowl from table 2 applies not only to Pacific flyway hunters in 1968 but to all hunters. Further assume that wetlands in the Canadian program yield 2.7 waterfowl per acre, as was estimated for the prairie wetlands purchased in the United States, and that breeding grounds are the sole limiting land resource in producing and sustaining waterfowl populations.

Invoke a steady state and equation (4-11),

$$X^* = I^* - cW^*$$

and further assume that c has a value of 0.24, implying values for S_1 and S_2 of 0.95 and 0.84 respectively. Then of the 2.7 waterfowl, hunters may shoot 2.05 (all but 24 percent) and yet maintain the breeder population at its present level. Assume 80 percent of the 2.05 birds shot are bagged, or 1.64. Multiply the 1.64 bagged waterfowl by the value \$3.29 to obtain the value of a wetland acre in waterfowl production, namely, \$5.40. This compares favorably to the \$4.73 per acre figure mentioned previously as the average annual payment for recently acquired easement land in the prairies as a whole, but does not compare favorably with the Alberta figure of \$5.60.[3] Considering the crudity of this exercise, no implication is made that the Alberta purchases are economically unjustifiable.

Now let us apply a discount rate to the \$5.40 figure of the value of a wetland acre in waterfowl production. At 6 percent, an asset yielding \$5.40

[1] The information is derived from the leaflet, "Dollars from Wetlands" (Ottawa: Canadian Wildlife Service, no date), and from A. S. Goodman, Biologist, Canadian Wildlife Service, in a personal communication dated October 7, 1969.

[2] Information from Burton W. Rounds, Wetlands Program Coordinator, U.S. Bureau of Sport Fisheries and Wildlife, in a personal communication dated August 8, 1969.

[3] The exchange rate differential between the currencies of the two countries is ignored.

in perpetuity would have a present value of $90; at 10 percent, of $54.[4] At the lower discount rate, purchase of U.S. breeding grounds can be justified through the entire range from $25 to $85 per acre.

This exercise clearly indicates the usefulness of a waterfowl valuation figure. Used in conjunction with an annual waterfowl production figure, adequate natural mortality and crippling loss estimates, and an alternate use value, any specific parcel of wetlands, large or small, may be rationally evaluated. The method does have inadequacies, as does the material developed in the last chapter. We turn to an elaboration of these problems and some solutions in the remainder of this chapter.

SHORTCOMINGS OF EARLIER APPROACHES

The policy that achieves maximum physical sustained yield of the variety discussed in chapter 4 is of limited value. It assumes implicitly that man cannot or does not control the amount of breeding habitat. However, should the number of wetlands be considered a policy variable, then there is in principle an uncountable number of maximum physical sustained yield solutions, one for each physical value of wetlands. A unique policy that picks the size of breeding population and the level of breeding environment that together maximize physical sustained yield cannot be found. If one asks how many ducks there should be or how large a harvest should be, the answer can only be found by choosing a higher level goal or by adding constraints that limit the number of permissible ponds.

Even if there are no feasible means of determining and controlling the desired level of wetlands, the goal of maximum physical sustained yield still is of limited value. More is not always better than less if this phrase is correctly interpreted. When the problem is seen through the eyes of hunters, they must decide whether it is better, more satisfying, to kill periodically the maximum physical sustained yield, or to kill more than this amount today by drawing down the breeding stock and enjoying lower seasonal harvest in future years. To argue that hunters prefer the first harvest pattern to the second is to reach conclusions by making an assumption about hunters' preferences marginally and intertemporally.

This concept of changing the breeding stock by changing kill essentially involves viewing waterfowl as having value as a capital good—a good used to make more goods in the future. The flaw in the cost–benefit analysis is now obvious. The waterfowl valuation figure used is derived from value in

[4] Economists have been arguing about the "correct" social rate of discount for years. For a bibliography on the subject, see William J. Baumol, "On the Social Rate of Discount," *American Economic Review*, vol. 58 (September 1968), p. 802.

consumption, and in no direct way is value in the production process taken into consideration.

Setting forth an alternative of greater scope and power than the treatments given here and in chapter 4 requires more sophisticated tools than those used so far. Undoubtedly such alternative treatment should consider wetlands a policy variable, for additional wetlands, by increasing the waterfowl population, can provide a greater annual harvest which has additional value to hunters. However, these benefits must be matched against the opportunity cost of the additional wetlands, that is, the net benefits forgone when cropland is taken out of production.

In the following discussion persons charged with managing the mallard population are assumed to seek that level of mallard population and supporting breeding habitat that maximizes the hunters' net willingness to pay for the hunting service through time. In this chapter we define net willingness to pay as the hunter's valuation of his activity, less his costs, and less the cost of obtaining additional wetlands.

Two basic models will be presented. The first model, continental in scope, treats the problem within the framework of continuous time using optimal control techniques. Numerical estimates of the parameters and functions provide an empirical counterpart to the analytical model. Optimal solution values for subregions in the North American continent are illustrated. The second continental model, actually a variant of the first, views the world as composed of discrete decision periods. We conclude approximately that in the long run, the manager should rent wetlands until the added cost of more wetlands is just offset by the value hunters place on the waterfowl produced on those wetlands. Given the desired number of wetlands, the best stock of waterfowl is that for which the (marginal) productivity of waterfowl, net of natural mortality, equals the rate of discount adopted by the management. Each hunter should be permitted to kill waterfowl until the value of an additional waterfowl to him equals the future value that the waterfowl could have contributed had it been left in its natural state. This rule is just like the savings–consumption decision many of us make periodically. Income is spent until the satisfaction obtained from an additional dollar (waterfowl) spent on contemporary goods and services is equated with the satisfaction a saved dollar and its yield provide in the future.

THE ANALYTICAL OPTIMAL CONTROL MODEL

The net value of hunting during a season is the product of N hunters and their individual valuation functions discussed in chapter 3

$$NV(D, Y, S, E) \qquad (5\text{-}1)$$

Note that the time subscripts have been dropped. The second ingredient of the story is a pond cost function noted by

$$C(P), \frac{\partial C}{\partial P} \equiv C_P > 0 \text{ for } P \geq \bar{P}, C_P = 0 \text{ for } P < \bar{P},$$
$$C_{PP} > 0 \text{ for } P \geq \bar{P} \text{ and } \lim_{P \to \infty} C(P) = \infty \quad (5\text{-}2)$$

In this scenario a given amount of wetlands, $\bar{P}$, is assumed to be freely available because the land has no alternative productive use. C_P is the marginal cost of ponds, expressed as an annual rental rate. It is therefore the annual easement value of the pond.

Specification of the value and cost functions alone is not sufficient, for waterfowl managers cannot act as free economic spirits with their enthusiasm for correct policy decisions tempered only by dollar comparisons. They must also recognize that the mallard population at any time is governed by the following differential equation:

$$\dot{W} = -W + S_2(I + S_1W - C_3ND) \quad (5\text{-}3)$$

This expression is equivalent to the population equation in chapter 4, with the following exceptions. The framework of the analysis has shifted from discrete to continuous time. The change simplifies the notation without loss of content although the reader should bear in mind that all variables are dated implicitly. A third parameter, C_3, has been added in explicit recognition that not all birds killed are bagged birds. Total hunter (N)-related physical losses to the system, including crippling losses and unretrieved kill, are estimated to be 1.25 times reported bag.[5] In review, reading from left to right within the parentheses, the young (I) plus the surviving breeders (S_1W) make up the fall flight from which the total kill (C_3ND) is subtracted to obtain the wintering population. Some fraction (S_2) of the wintering population survives to return to the northern breeding grounds in the spring. If this number is greater than the original number of breeders (W), clearly there has been a positive change; that is, the time derivative, dW/dt or $\dot{W}$, is positive.[6]

Since the agent in charge of maximization has a many-period problem, he must pick a discount rate, r, which describes the tradeoff between present and future valuable goods and services. Then, with knowledge of

[5] The expression for total hunter kill in this section (C_3ND) corresponds to the shorthand expression X in earlier chapters.

[6] The time derivative of any variable will be denoted by a dot over the variable.

the existing waterfowl population, $W(O)$, the problem is to find the paths of P and D which maximize

$$\int_0^\infty [NV(D, Y, S, E) - C(P)]e^{-rt}\, dt \tag{5-4}$$

subject to (5-3), $D > 0$, $P > 0$, and the assumption that $NV - C$ is strictly concave.[7] In this formulation each individual is given a welfare weight of one.

Formally, the expression to be maximized is

$$H = \{NV(D, Y, S, E) - C(P) - \lambda[W - S_2(I + S_1W - C_3ND)]\}e^{-rt} \tag{5-5}$$

where λ is an auxiliary variable. The functions have been specified in such a way that a solution, and therefore an optimal management program, exists.[8] The important necessary conditions for an interior maximum of H are

$$V_D - \lambda S_2C_3 = 0 \tag{5-6a}$$

$$-C_P + \lambda S_2I_P = 0 \tag{5-6b}$$

$$-\lambda(1 - S_2I_W - S_1S_2) = -\dot{\lambda} + r\lambda \tag{5-6c}$$

The transversality condition is

$$\lim_{t\to\infty} \lambda e^{-rt}W_t = 0$$

The imputed value of a marginal change in the waterfowl population is the economic interpretation given to λ. One can think of waterfowl in general and mallards in particular as capital that combines with ponds (labor) to produce young (output). Verbally stated, expression (5-6a) says that hunters should kill up to the point where their marginal valuation just equals the economic loss to the system if capital is reduced by a unit. Since a fraction of that mallard population would have died naturally,

[7] In reality, knowledge about the variables Y, S, and E is also needed. Because hunters' income, seasons hunted, and costs will change through time, estimates of these changes would have to be made if actual policies are to be formulated on the basis of this model. In this paper the empirical illustrations follow from the use of mean sample values for Y, S, E. This is a substantive simplifying assumption.

[8] One of the best treatments of optimal control problems in economics may be found in Kenneth Arrow and Mordecai Kurz, *Public Investment: The Rate of Return and Optimal Fiscal Policy* (Baltimore: Johns Hopkins Press, for Resources for the Future, 1970).

the hunter should only pay $S_2\lambda$, where $S_2\lambda$ is less than λ. But because he cripples some birds, his charge must be multiplied by C_3.

In this formulation, the manager can call out a charge $\lambda S_2 C_3$ and let each hunter bag birds until his marginal valuation equals the charge. Alternatively, he can set the bag limit for *each* hunter at the appropriate physical level. Although the latter policy is an optimal one within the bounds of this model, it is not a sensible one in reality. Information costs differ with respect to the two policies, yet this fact has not been included in the formal analysis. The charge policy requires the enforcing agent, in some instances the game warden, to verify the number of bagged waterfowl per hunter. Under the quantity rationing policy, the enforcing agent has to verify the number of bagged waterfowl and the marginal valuation function for each hunter encountered. This last piece of information is likely to be very expensive to obtain.

Expression (5-6b) gives the standard economic results that factors ought to be hired up to the point where the price of the factor C_P equals the value of its marginal product.

It is interpretively helpful to rewrite (5-6c) as

$$\lambda g = r\lambda - \dot{\lambda} \tag{5-7}$$

where $g = S_2 I_W + S_1 S_2 - 1$ and g is interpreted as the own net (percentage) rate of return on the natural resource. In a stationary state, when $\dot{\lambda} = 0$, equation (5-7) says, quite simply, that the best natural resource stock level is that for which the rate of discount equals the marginal (percentage) rate of growth of the natural resource with respect to the level of the resource stock. If one kind of capital is less productive than another kind at the margin, we should consume the less productive capital (waterfowl) and get satisfaction in so doing, until the marginal return on waterfowl capital equals the marginal return on capital elsewhere in the economy. When the stock of waterfowl is smaller than optimal, the marginal price of W is high and falls through time to its steady-state value. Hence, when g is greater than r, W is less than W^*,[9] the optimal value, and $\dot{\lambda}/\lambda < 0$. As is true for most types of capital, the marginal productivity of waterfowl is high for "small" amounts of W because time has built this survival mechanism into most animals. This explanation makes sense to an economist. In biological terms, the mallards exhibit compensatory mortality, as discussed in chapter 4.

[9] The asterisk refers to optimal steady-state values because there is only one optimal steady state in this section, compared to multiple maxima in the previous chapters.

EMPIRICAL RESULTS FOR THE OPTIMAL CONTROL MODEL

The transition from the analytical model to empirical results encounters two problem areas, both of which require discussion at the outset.

Ponds

Ponds must be well defined in terms of productivity, and ideally the cost of ponds should be estimated carefully. Our analysis is inadequate on both counts. In the empirical model, all wetlands are assumed to be homogeneous in productivity of both waterfowl and grain crops. In reality, ponds vary in their ability to support waterfowl just as the dry depressions vary in their ability to support agricultural crop growth.[10]

Since data on continental wetlands are not available for a sufficient number of years to permit parameterization, Canadian prairie pond data have been used as a proxy in the physical production functions. Our statistical results indicate that the proxy variable performs well, at least in the neighborhood of historical observations. But should the model call for the creation of additional wetlands in the Canadian prairie provinces, implicitly wetlands are also increased elsewhere on the North American continent. Ideally, our system would explicitly introduce a factor of proportion between prairie ponds and ponds elsewhere, and would consider the opportunity costs both of land now nurturing ponds that are endangered by drainage and of land upon which ponds may be created. In short, if ponds obtained through easement are less productive than wetlands freely available, or if the policy prescriptions emerging from this model call for a substantial increase in the number of ponds, the results will be biased. The direction of the bias is not difficult to discover but its magnitude is not easily ascertained owing to the inadequacy of data.

Estimates of pond costs are also problematical. The Canadian government's support program for wetland purchases and easements provides suggestive data. As stated earlier, average annual payments made on recently acquired easement land are $5.60 per acre for land in Alberta, the area that provides much of the waterfowl shot by Pacific flyway hunters in the United States.[11] According to administration of the Canadian easement program, there are about 0.85 acres in a pond and its supporting

[10] In this particular respect, the exercise in the first part of this chapter is superior.

[11] The origin of waterfowl bagged in various flyways is estimated from banding data. Approximately one-half "of the recoveries within the United States from bandings in Alberta were in the Pacific Flyway. . . ." U.S. Bureau of Sport Fisheries and Wildlife, *Distribution of Recoveries from Bandings of Ducklings*, Special Scientific Report—Wildlife No. 89 (Washington, D.C., December 1964), p. 3.

wetlands. Thus, ponds have an easement cost of $4.76 per pond, on the basis of recent experience. How many existing or threatened wetlands could be obtained at this price is not known.

Alternatively, one can try to estimate the opportunity cost of marginally productive lands for which wetlands are the best alternative to agriculture. A recent linear programming study drawing on representative farm or crop budgets, said to characterize the Alberta region, produced annual estimates in the neighborhood of $17 per wetland acre.[12] If this number is accurate, it establishes a sort of upper bound on the value of marginal wetlands. In other words, at the margin, the value of land drained for agricultural purposes less the annual drainage investment should approximate the value of land in its next best nonagricultural use. The figure would be an appropriate one to use if the investment cost just referred to is approximately equal to the cost of purchasing the implicit ponds which must be added outside the prairie provinces.

The costs mentioned thus far refer only to maintaining the permanency of existing wetlands, but new wetlands can be created. According to Ducks Unlimited:

> Simple earth dams on the outlet of marshes store additional water to outlast periods of drought. Restriction of evaporation area is also effective. Dikes built across shallow marshes or lakes which are frequently dry, confine the water to half or less of the former area. There are also many large shallow prairie lakes too salty to be of much value to breeding waterfowl. Often inflowing creeks enter these lakes through shallow bays of lower salinity. A dam across the neck of the bay where it joins the lake proper develops the bay into good duck-breeding habitat. An inflowing creek can also be diverted through a series of natural basins en route to the lake. Creeks and other flowages, less well defined, can also be diverted into marshes and shallow lakes which lack sufficient water. New marshes can be created by diverting water to natural depressions in the terrain through the use of dams and other controls. Many fine areas have been developed in southern Alberta by using waste irrigation waters.[13]

Ducks Unlimited estimates that its recent projects cost it slightly more than $1 per acre. The low figure is largely attributed to the fact that it buys no land.[14]

In summary, estimates of the costs of a pond vary from about $1 to $17. Fortunately, the choice of an estimate is considerably narrowed. In all the

[12] Prairie Agri-Management Consultants, Ltd., *Sensitivity Analysis of the Estimated Opportunity Cost of Maintaining Wetlands*, reports I and II, prepared for the Canadian Wildlife Service, February 20, 1970.

[13] Ducks Unlimited (Canada), "Information: A History of Ducks Unlimited," p. 5.

[14] Personal communication from Ducks Unlimited (Canada), January 22, 1971.

cases described below, a pond cost of $4.76 produces an optimum number of ponds that ranges from five to more than twenty times the historical average, indicating that new ponds should be created. In our judgment, Ducks Unlimited does not have sufficient moral suasion to coax that much land out of current production, and it is doubtful that for the preservation price of $4.76 per acre, the easement program could attract the 7 to 36 million acres of ponds called for in the solution to the optimum steady-state program. Therefore, cases assuming values of $17 per pond and $12 per pond are illustrated. The latter figure was chosen arbitrarily. While the actual marginal cost of a pond may be neither $12 or $17, the range indicates the sensitivity of the optimum solution values to changes of this order of magnitude.

Joining the Physical and Valuation Functions

Regrettably, the scope of this study was restricted by intellectual myopia combined with the lack of unlimited funds. The sample space for the waterfowl valuation function is the Pacific flyway while the physical functions refer to a waterfowl population continental in scope. Joining the economic and physical parts requires either that the economic relations be broadened or that the physical functions be reduced from the continent to the Pacific flyway. We have chosen the first option on grounds of simplicity. Decomposition of the aggregate physical function is a relatively simple analytical task only in the case of the constant returns to scale (*CRS*) production function that used May pond data. If nature yields constant returns to scale, that is, if an equal increase in both breeders and ponds produces the same proportionate increase in young, then a proportionate expansion of breeders and ponds in regions such as Alberta and Saskatchewan, which produce mallards for the Pacific flyway, neither decreases nor increases the productivity of mallards elsewhere on the continent. For the other production functions studied, intellectual experiments involving decomposition of the nonlinear physical functions were not rewarding. To obtain empirical results for the continent, we make the assumption that all hunters in Canada and all other hunters in the United States (Central, Mississippi, and Atlantic flyways) have the same tastes, average income, average seasons hunted, and average costs as those sampled in the Pacific flyway. Shreds of evidence suggest that this may be a strong assumption, but there is no alternative in the absence of economic data for the other regions within the continent.

Empirical Results

The solution system of equations (5-3), (5-6a), (5-6b), (5-6c), the regression results, and the parameter values enable the hypothetical manager of water-

fowl to determine the optimal breeders, ponds, and kill under steady-state conditions for the multiplicative (*MU*), constant returns to scale (*CRS*), and Beverton–Holt (*BH*) production functions.[15] A discount rate of 8 percent was arbitrarily chosen.[16] Readers will note that the mallard data are used as a substitute for all species of waterfowl because no reliable data are available for the others.[17] Table 10 summarizes the empirical results for a pond cost level of $12 (cases *MU*12, *CRS*12, *BH*12) and $17 (cases *MU*17, *CRS*17, *BH*17).

Before discussing the results in table 10, it is natural to draw a comparison between maximum physical sustained yield solutions discussed earlier and the optimum economic solutions. There are dramatic differences between the two models when ponds are worth $12 each, *MU*12, and the physical maximum for the multiplicative case (4-19) derived for the average number of ponds during the dry cycle years of 1958–69 (table 7). The economic optimum calls for about 5 million more breeders, about 5 million more ponds, and of course, the mallard harvest is very much greater, by more than 5 million. The economic optimum number of breeders can be compared qualitatively with the biological optimum, each evaluated at the *economic* optimum number of ponds.[18] The qualitative results are self-evident. The biological optimum calls for relatively more waterfowl which combine with the same number of ponds to produce a higher annual kill and a lower marginal value.[19]

How does the recent historical average compare with one economic optimum, *MU*12? From 1961 to 1968 the estimated average breeders and July ponds were 7.8 million and 1.3 million respectively, while the *MU*12 optimum calls for 15.2 million breeders and 6.3 million ponds, a sizable difference indeed. Historically, the continental mallard harvest averaged about 4 million whereas the optimum produces about 12 million, again a

[15] The parameter values chosen are $S = 0.95$, $S_2 = 0.84$, $C_3 = 1.25$, the number of hunters in the Pacific flyway = 0.279 million and the total number of hunters = 1.715 million. The last two figures were estimated from U.S. Bureau of Sport Fisheries and Wildlife, *Waterfowl Status Report*, Annual Series (Washington, D.C.).

[16] A lower discount rate typically means that an economy accumulates a greater store of wealth including, in this case, more waterfowl. More waterfowl produce a larger stationary yield, implying a lower stationary marginal value for the waterfowl harvested. Moreover, a lower discount rate for a given pond cost induces a change in the factor proportions, increasing the ratio of W to P.

[17] Historically, mallards (M) constitute 30 percent of the hunters' average kill (D), or $D = 3.33M$, *Waterfowl Status Report*, Annual Series.

[18] We emphasize that biologists could not determine the economic optimum number of ponds without conducting the economic analysis.

[19] Maximum physical sustained yield implies an own rate of return equal to zero. For this to occur, the waterfowl population must be increased beyond the level producing a positive own rate of return.

Table 10. Illustrative Stationary Economic Optimal Values

	Historical values 1961–68[a]	Cost of pond = \$12			Cost of pond = \$17		
		*MU*12	*CRS*12	*BH*12	*MU*17	*CRS*17	*BH*17
W Breeders (millions)	7.8	15.2	17.2	12.1	11.4	9.5	9.5
P July ponds (millions)	1.3	6.3	7.5	2.9	4.0	4.2	2.0
λ Marginal value of a waterfowl (dollars)		3.10	2.40	4.00	3.70	3.40	4.65
Total continental mallard kill (millions)	3.7	12.2	19.2	8.1	9.1	10.6	6.2
Mallard kill per hunter, Pacific flyway	3.5[b]	7.1	11.2	4.7	5.3	6.2	3.6
Total mallard kill, Pacific flyway (millions)	0.9	2.0	3.1	1.3	1.5	1.7	1.0
Total kill per hunter, Pacific flyway	11[b]	24	37	16	18	21	12
Total kill, Pacific flyway (millions)	3.6[b]	6.6	10.4	4.4	4.9	5.7	3.3

[a] Data computed from U.S. Bureau of Sport Fisheries and Wildlife Service, *Waterfowl Status Report*, Annual Series.
[b] Average for years 1965–69.
Note: Assumptions were as follows: discount rate = 8%; $S_1 = 0.95$, $S_2 = 0.84$; $C_3 = 1.25$; number of hunters in the Pacific flyway = 0.279 million; total number of hunters = 1.715 million.
MU refers to the multiplicative production function using July pond data, *CRS* refers to the multiplicative production function using May pond data and yields constant returns to scale, and *BH* refers to the modified Beverton and Holt relationship. As a mnemonic device, marginal pond cost in dollars follows the production function symbol.

very sizable difference. All cases prescribe more breeders, more ponds, and greater kill than recent experience has provided.

One can see quite clearly the quantitative corroboration of an anticipated conclusion reached earlier. Recognition of the present economic productivity of ponds calls for an increase in their number. Such an increase is coupled with a lesser relative increase in breeders, and the factor proportion W/P falls. The average historical W/P is 6.0, while $(W/P)^*$ ranges between 2.3 and 4.8, depending upon the particular case. Past mallard harvests averaged under 4 million, whereas the optimum harvest ranges from 6 to 12 million mallards for the continent.

The results for the Pacific flyway in each case are obtained by reducing the total continental kill or total mallard kill by the fraction of Pacific flyway hunters to total hunters.[20] From the Pacific flyway hunter's point of view, optimum management of waterfowl does not decrease his seasonal kill. Indeed, if wetlands can be purchased for $12 each, his kill rises from 11 waterfowl per season to between 16 and 37, depending on the case chosen. Of course, in addition to his present hunting expenditures he now has to pay from $2.40 to $4 each for killing privileges if pond costs are $12.

Although the hunter's seasonal kill now is higher, he is not better off if he in fact must pay for the right to bag birds. Under this policy his net gain is $V(D^*) - D^*V_{D^*}$, which must be compared to the value of his kill in the absence of the proposed management policy, $V(D)$. It is easy to show that

$$[V(D^*) - D^*V_{D^*}] - V(D) < 0 \tag{5-8}$$

From (3-3) and (3-4)

$$V(D^*) - D^*V_{D^*} = (1 - q_4)V(D^*) \tag{5-9}$$

Substitution into the above expression yields

$$(1 - q_4) < \frac{V(D)}{V(D^*)} \tag{5-10}$$

which can be written explicitly, using (3-4) evaluated at the mean values of the exogenous variables and the estimated value of $q_4 = 0.409$, as

$$0.27 < \frac{D}{D^*} \tag{5-11}$$

Since the largest estimated ratio of D/D^* is 0.30, the hunter is not better off under this system. In a decentralized system of decision making, the owners of ponds clearly are the gainers, as now they are rewarded for maintaining ponds or building new ones.

Having compared our economic approach with recent history, let us discuss the alternative economic optimizing formulations, beginning with *CRS*17 and *MU*17. These two cases are in closest correspondence with differences between W, P, λ, and kill, which are no greater than 20 percent. When ponds cost $12, the variation between *MU*12 and *CRS*12 values is much greater (compared to the $17 cases), and for sound reasons. A fall in the price of ponds from $17 to $12 signals expansion of the system. The expansion encounters diminishing returns in the instance of *MU*12, so neither ponds nor waterfowl expand as much as in the constant returns to

[20] In order to avoid possible confusion, it should be pointed out that earlier dismission focused on the legitimacy of maximizing a fraction of the system. Having solved the aggregate system, we now are exhibiting the results for a portion of the system.

scale case. This is the principal explanation for harvest under *CRS*12 being more than 50 percent greater than that under *MU*12.

We find it difficult to provide a profound explanation for the difference between the *MU* and *BH* cases. The optimum mallard population is not greatly different for a given pond cost, but under the *BH* physical characterization, fewer ponds are prescribed and annual harvest is substantially lower. According to tables 8 and 9, *BH* is more productive for a given waterfowl population as long as it is above 6 million, but the comparative advantage of *BH* falls as the number of ponds increases. In other words, for a large-scale system—when ponds cost under $17—the *BH* characterization is relatively unproductive. However, a sufficiently high rate of discount and pond cost yields a harvest in the *BH* model larger than in rival models.

OPTIMAL SOLUTION VALUES FOR SUBREGIONS IN THE NORTH AMERICAN CONTINENT

The empirical results thus far regard all hunters as having the same average valuation function for waterfowl. It is of some interest to test the assumption of hunter homogeneity. This was accomplished by solving the optimal control model with eight subregions.[21] The Pacific flyway was divided into seven regions: Arizona and New Mexico; California; Idaho and Montana; Nevada; Oregon; Utah, Colorado, and Wyoming; and Washington. The final region was the rest of the North American continent.

Table 11 contains the hunter population in each region as well as the mean values for the arguments used in each regional valuation function. It will be noted that the rest of the continent is given the mean values obtained from our sample population.

The results of the optimal allocation can be anticipated for some regions. The variables *Y*, *S*, and *E* are positive demand shifters. According to table 11, California hunters have values for *Y*, *S*, and *E* each greater than the respective aggregate mean values; thus their optimal bag of waterfowl or mallards should be greater than the regional average. This is illustrated in figure 12. Notice in figure 12 that hunters pay the same charge regardless of their location. This makes sense on a moment's reflection. According to the competitive theory of the firm, the owner of a firm sells his product to everyone for the same price if the marginal cost of producing a good or service is the same, regardless of the destination of the good. Since the

[21] The Hamiltonian is

$$H = \{ \sum N_i V_i(D, Y, S, E) - C(P) - \lambda[W - S_2(I + S_1 W - C_3 \sum N_i D_i)]\} e^{-rt}$$

where the subscript $i = 1...8$, denotes the region.

Table 11. Regional Mean Values for Waterfowl Hunters' Income, Seasons Hunted, and Costs

Region	Income/ hunter (thousands of dollars)	Seasons hunted S_i	Annual cost (dollars) C_i	Hunters (millions) N_i
Arizona and New Mexico	10.7	15.2	295	0.007
California	13.8	16.2	330	0.121
Idaho and Montana	10.2	16.1	255	0.035
Nevada	11.2	16.4	312	0.009
Oregon	12.0	17.7	343	0.032
Utah, Colorado, and Wyoming	9.8	14.0	345	0.030
Washington	10.4	16.2	309	0.048
Rest of continent	12.1	16.2	322	1.434
Mean value	12.1	16.2	322	—
Total				1.719

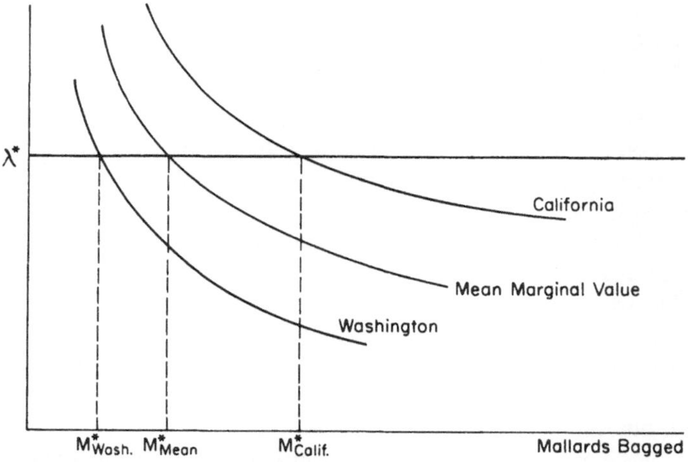

Figure 12. Relative optimal bag levels for selected states.

fictional producer of birds incurs no extra cost if the birds fly to Washington or to California, his marginal cost is the same. Economists might be tempted to argue that waterfowl that migrate longer distances have a greater natural rate of mortality—the capital depreciates more rapidly. In this case the true shadow price, $(\lambda S_2 C_3)$, would differ among regions because of regional differences in S_2.

The Hamiltonian is

$$H = \{\Sigma N_i V_i(D, Y, S, E) - C(P) - \lambda[W - S_2(I + S_1 W - C_3 \Sigma N_i D_i)]\}e^{-rt}, \quad i = 1, \ldots 8 \qquad (5\text{-}12)$$

where the subscript i denotes the region. But our previous discussion of estimated values of S_2 indicates that biological knowledge presently is too

imprecise to permit regional estimates of S_2. Additionally, waterfowl that fly greater distances may have, through natural selection, the same or greater rates of survival.

From table 11 we also can conclude that the mallard bag level for Arizona, New Mexico, Idaho, Montana, and Washington will be below the mean because hunters in these states have values Y, S, and C, the same as, or lower than the respective mean values.

Table 12 illustrates the optimal level of mallard kill by region for four cases, *MU*12, *CRS*12, *MU*17, and *CRS*17. For any given case the maximum difference in bagged mallards is not more than 3.2 per season (*CRS*12). The difference diminishes as the pond cost increases. California hunters rank first in amount bagged while Idaho and Montana rank last. The optimal values for kill in the "other" category (which includes 82 percent of the hunters) is insignificantly different, case by case, from the optimal values presented in table 10.

On the basis of earlier empirical results, we concluded that the lack of markets gives rise to a substantial underinvestment in ducks. What about intra-Pacific flyway externalities? How well do managers allocate waterfowl among the principal regions in the Pacific flyway? Comparing the ideal share of waterfowl to the actual share based on the questionnaire

Table 12. Optimal Level of Mallard Kill per Hunter by Regions

		Cost of pond = $12		Cost of pond = $17	
Region	Rank	*MU*12	*CRS*12	*MU*17	*CRS*17
Arizona and New Mexico	6	6.3	9.8	4.7	5.5
California	1	8.0	12.5	6.0	6.9
Idaho and Montana	8	6.0	9.3	4.4	5.2
Nevada	4	6.8	10.6	5.0	5.8
Oregon	2	7.5	11.6	5.6	6.4
Utah, Colorado, and Wyoming	7	6.0	9.4	4.5	5.2
Washington	5	6.4	9.9	4.7	5.5
Rest of continent	3	7.2	11.2	5.4	6.2

Table 13. Ideal and Actual Share of Kill for the Regions of the Pacific Flyway

Region	Ideal share (percent)	Recent actual share (percent)
Arizona and New Mexico	2	2
California	49	50
Idaho and Montana	10	10
Nevada	3	3
Oregon	12	14
Utah, Colorado, and Wyoming	9	8
Washington	15	14
Total	100	100

data produces astonishing results, we think. The percentages presented in table 13 indicate that the physical policies of season length and bag limits established for each state result in actual shares that are virtually identical to the ideal shares. The greatest difference is 2 percent (Oregon). We conclude that the intra-Pacific flyway externalities are not substantial, when observed within the confines of our model. However, we hasten to add that these values should be viewed as illustrative of what a model of this detail can do. They should not be regarded as definitive estimates.

A DISCRETE-TIME, INFINITE-HORIZON MODEL

Readers of earlier drafts of this work have argued that a model incorporating discrete time is more appropriate than a continuous time approach because both the hunting season and the period of production occur during fractional parts of the calendar year. The analytical results are identical in either case, apart from an inconsequential appreciation factor. It has also been argued that our formulation of the control theory problem is incorrect because the differential equation (5-3) for W is wrong. The first model, embracing (5-3), in fact is correct, as will be verified by the dynamic programming approach. In our judgment the economic interpretation of the continuous time approach is much easier to understand because of its notational simplicity. We therefore prefer it for this reason, but readily admit that it does take some liberties with reality since man's harvest and nature's birth process are periodic. On the other hand, a faithful characterization of the bioeconomic story would involve using a mixed difference–differential equation since natural mortality is continuous. Readers uninterested in pursuing refinements, including the development of a phase diagram characterizing the evolution of the system out of the steady state, should turn to the next chapter.

The first familiar ingredients of the problem are a hunter's net valuation function in period t,

$$V(D_t, Y_t, S_t, E_t) \tag{5-13}$$

where the arguments of V have the same economic interpretation and the following conditions on V must hold:

$$V_{D_t}, V_{Y_t}, V_{S_t}, V_{E_t} > 0$$

$$V_{D_tD_t}, V_{Y_tY_t}, V_{S_tS_t}, V_{E_tE_t} < 0$$

and

$$\lim_{D_t \to 0} V_{D_t} = \infty$$

Then there is the pond cost function

$$C(P_t) \tag{5-14}$$

where $C_{P_t} > 0$, $C_{P_tP_t} \geq 0$. Additionally $V - C$ is strictly concave. The difference equation describing the waterfowl population is

$$W_{t+1} = S_1S_2W_t + S_2f(W_t, P_t) - NS_2C_3D_{t+1} \tag{5-15}$$

where S_1 and S_2 are fractions of the population naturally surviving.[22] The annual production function for immatures $f(W_t, P_t)$ is assumed to be concave. The parameters N and C_3 have the same interpretation. Time periods have been defined to end at the time of the fall flight and not at the end of a calendar year.

The figurative waterfowl manager displaying a civic, altruistic heart has an infinite planning horizon, and seeks to find the optimal net economic value of the natural resource by maximizing the following Lagrange function:

$$L = \sum_{1}^{\infty} \theta^t\{NV(D_t) - C(P_t) - \lambda_t[W_t + NS_2C_3D_t - S_1S_2W_{t-1} - S_2f(W_{t-1}, P_{t-1})]\} \tag{5-16}$$

subject to $D_t \geq 0$, $P_t \geq 0$ and given $W_0 > 0$, $P_0 > 0$.[23] Additionally,[24]

$$\theta^t = \frac{1}{(1 + r)^t}, \qquad r < 1$$

For purposes of exposition it will be supposed that an optimal policy never involves a choice of a zero value for D_t, P_t, or W_t.[25] The necessary conditions for an interior maximum are

$$V_{D_t} - S_2C_3\lambda_t = 0 \tag{5-17}$$

$$-C_{P_t} + \theta S_2\lambda_{t+1}f_{P_t} = 0 \tag{5-18}$$

$$-\lambda_t + \theta\lambda_{t+1}(S_1S_2 + S_2f_{W_t}) = 0 \tag{5-19}$$

for all t, and the satisfaction of (5-15). These conditions together with the conditions on V, $V - C$, f, along with the assumptions of D_t, P_t, and $W_t > 0$, an assumption that

$$\lim_{t \to \infty} \lambda_t\theta^tW_t = 0$$

[22] Survival rates and survival fractions are given the same notation. The numerical difference between the two concepts is trivial in this instance.

[23] Individual variations in the arguments of $V(.)$ are submerged to avoid clutter. See note 7.

[24] Adding θ^t to the Lagrange multiplier, λ_t, does not affect maximization (Arrow and Kurz, *Public Investment*, pp. 46–47), but does facilitate the economic interpretation of the results below.

[25] This assumption permits us to omit the inequality signs in the necessary conditions. A more general solution to this problem is contained in O. Burt and R. Cummings, "Production and Investment in Natural Resource Industries," *American Economic Review*, vol. 60 (September 1970), pp. 576–590.

and the stronger assumption of stationarity are sufficient to ensure a maximum.[26]

The interpretation of (5-19) is readily understood in the stationary state when $\lambda_{t+1} = \lambda_t = \lambda$, for it simply says that the best natural resource stock level is that for which the rate of discount equals the rate of growth of the natural resource. Let $\Delta W = W_{t+1} - W_t$ and $S_2C_3 = C_4$. Then from (5-15)

$$\Delta W = (S_1S_2 - 1)W_t + S_2f(W_t, P_t) - NC_4D_{t+1} \qquad (5\text{-}20)$$

and the marginal rate of growth of the natural resource, $g(W_t, P_t)$

$$\frac{\partial \Delta W}{\partial W_t} \equiv g(W_t, P_t) = (S_1S_2 - 1) + S_2f_{W_t} \qquad (5\text{-}21)$$

Recalling that $\theta = 1/(1 + r)$, expression (5-19) can be written as

$$1 + r = S_1S_2 + S_2f_{W_t} \qquad (5\text{-}22)$$

Finally, substituting (5-21) into (5-22) yields the desired result

$$r = g(W^*, P^*)$$

where the asterisk refers to stationary values. This final result mirrors that obtained in the optimal control model, namely that the best natural resource stock level is that for which the discount rate equals the rate of growth of the natural resource.

Rearranging (5-18) in more familiar form,

$$\lambda_t S_2 f_{P_t} = (1 + r)C_{P_{t-1}}$$

produces the unsurprising conclusion that in every period ponds ought to be rented up to the point where the value of the marginal product of ponds equals the marginal cost of ponds ($C_{P_{t-1}}$), which must be appreciated by $(1 + r)$ because production costs occur one period prior to "sale" of the product. The periodic model differs analytically from the continuous model only by this appreciation factor. Actually the two models are even closer than this because we have been a bit too casual with time subscripts. In principle, the ponds can be rented in the spring and the birds harvested in the late fall for a lapse of less than a year. If the fraction of the year between the expenditures for ponds and the "sales" receipts is μ, then the correct appreciation factor is $1 + \mu r$ if r is thought of as the annual rate of interest. If $r = 8$ percent and $\mu = 3/4$, the real cost is 6 percent greater than C_{P_t}. Thus, the optimal values for the steady-state continuous model (table 11) equal those obtained in the discrete model if illustrative pond cost values of \$12 and \$17 are replaced by \$11.30 and

[26] Arrow and Kurz, *Public Investment*, p. 21.

\$16 respectively. The interpretation of (5-17) is exactly equivalent to that for (5-6a).

The phase diagram, figure 13, portrays the evolution of this system through time as it moves toward and reaches a stationary equilibrium.

The phase diagram is a geometric representation of what we know from good economic intuition or have derived mathematically. There are lots of combinations of bagged waterfowl (D_{t+1}) and waterfowl populations (W_t) for which the population does not change (any point on $\Delta W = 0$), or the level of bagged waterfowl does not change (any point on $\Delta D = 0$), but only one pair of values $D_{t+1}{}^*$, $W_t{}^*$, for which the population and bag level remain constant period after period. One can begin with a starting level of waterfowl and the solution for the difference equations here represented by the "feathered" line, which dictates what the bag should be each and every subsequent period, as well as the best level of ponds. But the behavior of ponds is hidden in this two-dimensional story. If the manager inherited W_0, he should permit D_1. If he mistakenly permitted $\tilde{D}_1$ and never tried to correct for the error, he would veer off to the northeast, as indicated by the arrows pointed north and east in that region, until hitting the $\Delta D = 0$ line, at which time he would choose values for D_{t+1} that send the system off on a southeast path. There is in this system only one road to Rome, so to speak. If one gets off that road and is foolish

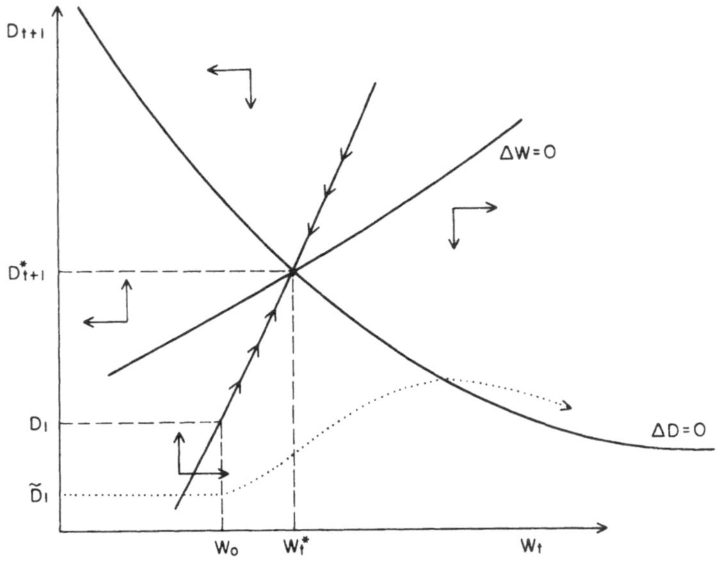

Figure 13. Phase diagram of the waterfowl system.

enough not to inquire about how to get back on (by solving a new maximization problem), then one cannot reach Rome, even fortuitously.

The proof of the motion of this system follows. From (5-17),

$$V_{D_{t+1}} - V_{D_t} = C_4(\lambda_{t+1} - \lambda_t) \tag{5-23}$$

Subtracting λ_t from (5-19), substituting (5-17) into (5-19) and the result into (5-23) gives

$$V_{D_{t+1}} - V_{D_t} = V_{D_t}\left[\frac{1}{\theta(S_1S_2 + S_2f_{W_t})} - 1\right] \tag{5-24}$$

From the explicit form of the Cobb–Douglas production function, $AW^\alpha P^\beta$, the marginal productivity of waterfowl is

$$f_{W_t} = \alpha A W_t^{\alpha-1} P_t^\beta \tag{5-25}$$

and the marginal productivity of ponds is

$$f_{P_t} = \beta A W_t^\alpha P_t^{\beta-1} \tag{5-26}$$

Equations (5-17), (5-18), and (5-26), after rearranging terms, yield

$$P_t^\beta = \left(\frac{\beta_5}{V_{D_{t+1}} W_t^\alpha A}\right)^{\frac{\beta}{\beta-1}} \tag{5-27}$$

where $\beta_5 = \dfrac{C_4\beta_4}{\theta S_2\beta} = \dfrac{C_3\beta_4}{\theta\beta}$,

and $C_{P_t} = \beta_4$, a constant.

When the result of substituting (5-27) into (5-25) is substituted into (5-24), the following expression is obtained:

$$V_{D_{t+1}} - V_{D_t}$$

$$= V_{D_t}\left[\theta S_1S_2 + \theta S_2\alpha A^{-\frac{1}{\beta-1}} W_t^{\frac{1-\alpha-\beta}{\beta-1}} \left(\frac{\beta_5}{V_{D_{t+1}}}\right)^{\frac{\beta}{\beta-1}}\right]^{-1} - V_{D_t} \tag{5-28}$$

Let

$$\phi(W_t, D_{t+1}) = \theta S_2\alpha A^{-\frac{1}{\beta-1}} W_t^{\frac{1-\alpha-\beta}{\beta-1}} \left(\frac{\beta_5}{V_{D_{t+1}}}\right)^{\frac{\beta}{\beta-1}}$$

which is positive when W_t and D_{t+1} are positive.

When $D_{t+1} - D_t \equiv \Delta D = 0$

$$\phi(W_t, D_{t+1}) + \theta S_1S_2 = 1 \tag{5-29}$$

and from (5-29)

$$\left.\frac{dD_{t+1}}{dW_t}\right|_{\Delta D=0} = \left(\frac{1-\alpha-\beta}{\beta}\right)\frac{V_{D_{t+1}}}{W_tV_{D_{t+1}D_{t+1}}} < 0$$

since $\alpha + \beta < 1$ and $V_{D_{t+1}D_{t+1}} < 0$ from the text.

From (5-15),

$$W_{t+1} - W_t = (S_1S_2 - 1)W_t + S_2f(W_t, P_t) - C_4ND_{t+1} \tag{5-30}$$

Substituting (5-27) into (5-30) yields explicitly

$$W_{t+1} - W_t = (S_1S_2 - 1)W_t + A^{-\frac{1}{\beta-1}}S_2W_t^{-\frac{\alpha}{\beta-1}}\left(\frac{V_{D_{t+1}}}{\beta_5}\right)^{-\frac{\beta}{\beta-1}} - C_4ND_{t+1} \tag{5-31}$$

When $W_{t+1} - W_t \equiv \Delta W = 0$

$$\left.\frac{dD_{t+1}}{dW_t}\right|_{\Delta W=0} = \frac{\left[(S_1S_2 - 1) - \dfrac{\alpha}{(\beta - 1)}\dfrac{\psi}{W_t}\right]}{\left[C_4N + \dfrac{\beta\psi}{(\beta - 1)}\dfrac{V_{D_{t+1}D_{t+1}}}{V_{D_{t+1}}}\right]} \tag{5-32}$$

where $\psi = \psi(W_t, D_{t+1})$ is the middle expression of the right-hand side of (5-31). The denominator of (5-32) is always positive.

We now show that when $D_{t+1} - D_t = 0$, the numerator of (5-32) must be positive. If this is not true, from (5-32)

$$(S_1S_2 - 1) - \frac{\alpha}{(\beta - 1)}\frac{\psi}{W_t} \leq 0 \tag{5-33}$$

But when $D_{t+1} - D_t = 0$, it follows from (5-28) that

$$\frac{\psi}{W} = \frac{1 - S_1S_2 + r}{\alpha} \tag{5-34}$$

Substitution of (5-24) into (5-33) yields

$$\frac{\beta(S_1S_2 - 1) - r}{\beta - 1} \leq 0 \tag{5-35}$$

which is a contradiction because $\beta < 1$ as is S_1S_2. Therefore

$$\left.\frac{dD_{t+1}}{dW_t}\right|_{\Delta W=0} > 0$$

whenever a stationary solution exists. Moreover, it can be shown that

$$\left.\frac{d\Delta W}{dD_{t+1}}\right|_{W_t \text{ constant}} < 0 \quad \text{and} \quad \left.\frac{d\Delta D}{dW_t}\right|_{D_{t+1} \text{ constant}} < 0$$

These facts indicate the motion of the system in figure 13. Finally, from (5-17) or (5-6a) it follows that as D_t increases over time, λ_t falls. The assertion made in the text is now proved.

chapter six Conclusion

The quotation from Pothier that opens this work implies difficulties in defining property rights with respect to wild creatures, including waterfowl. Of course, such difficulties often give rise to externalities, that is, transfers of value in which at least one entity is an involuntary participant. For instance, a Canadian farmer with a cut grain field does not voluntarily open it to feeding waterfowl, yet value is transferred nevertheless to hunters who bag the birds.

FARMER–HUNTER EXTERNALITIES

None of our analyses has taken crop depredation costs into account, and, we think, for sufficient reason. Although crop damage presumably varies directly with the size of the fall flight and with the size of population at the wintering habitat, present crude statistical evidence will not support this view. Crop damage in Canada is related most fundamentally to the date and length of the crop harvesting activity relative to the time when waterfowl begin migrating south. If the mechanism governing the time of fall flight in the mallards sends an immense population southward before grain crops have been cut or after the crops have been combined (i.e., threshed), damage will be minimal. If even a small waterfowl population

migrates soon after cutting, the crop damage will be severe.[1] With present knowledge, the time of departure–time of cutting effect swamps any population effect on crop damage magnitudes.

Farther south, feeding programs and changes in farm technology have brought about significant reductions in waterfowl depredation, at least in some areas. California provides an excellent example. Major waterfowl damage occurred in the 1940s and 1950s,[2] but measures taken since then have reduced the problem currently to minor proportions. New varieties of rice mature earlier and are so dense that the underlying water often cannot be seen from the air by birds, thus reducing damage to that grain crop. Watergrass (wild millet) now is planted in feeding areas of federal and state refuges. Waterfowl find this grain highly palatable, seeming to prefer it to agricultural grains, and it has the added advantage of producing a volunteer crop for several years after the original sowing. If the agricultural harvesting season is late during a particular year, the opening of refuges to hunting is delayed to avoid scattering the birds congregated there.[3]

California has instituted another pioneering, if somewhat more esoteric, program in the area of crop depredation by waterfowl. The approach rests on what may be called the excremental technique for population estimation. As reported by Earl R. Biehn,[4] a number of ducks were penned and given a suitable diet. The average number of droppings per bird per day was observed, namely 60. This figure, together with appropriate and careful scrutiny of any injured agricultural field, yields the number of duck-days damage.

Even with the aid of such promising research efforts by others, we are unable to show that incremental or decremental changes in the waterfowl population affect the size of crop depredation; our omission of crop damage as a function of waterfowl population is consistent with present knowledge.[5]

Another externality from farmer to hunter arises from present institutions whereby farmers are not paid for the waterfowl "crop" produced

[1] Renewable Resources Consulting Services Limited, "A Study of Waterfowl Damage to Commercial Grain Crops in Alberta," (Edmonton: 1969), pp. 51–52, 57, 59, 69.

[2] Earl R. Biehn, *Crop Damage by Wildlife in California*, Game Bulletin No. 5 (Sacramento: California Department of Fish and Game, 1951).

[3] Frank Kozlik, of the California Department of Fish and Game, in a telephone conversation, August 15, 1969.

[4] *Crop Damage*, pp. 50–53.

[5] If crop damage is a function of population, the economic consequences are: (1) the optimal size of the population will be lower; (2) the marginal value of a bagged duck will be higher, so the number of ducks killed per season will be lower; and (3) the breeder/pond factor proportion will be lower than the results described in chapter 5.

on their land. Clearly, if one believes that those who benefit from a good should pay for it, waterfowl hunters in an ideal world would bear the full costs of their sport. This would include costs of wetlands purchase (a program partially correcting for present inequities to farmers), refuge upkeep, and general administrative costs of the waterfowl program.

Also, in most economists' ideal world there would be no farm crop subsidies. One of the major conclusions of Jon H. Goldstein's study of U.S. Midwest wetlands was that drainage of much of the land would not be worthwhile in the absence of agricultural drainage and price supports.[6] If agricultural subsidies are considered an income transfer to farmers that is endorsed by the general public, then an argument can be made that the general public should pay that portion of wetlands purchase price attributable to increased value resulting from the subsidies. The public does not so pay, as governmental wetlands purchases are financed from the sale of waterfowl hunting permit stamps, but the public does pay some general administrative costs of the waterfowl program. Hence there is some balancing out.

OTHER EXTERNALITIES

An additional externality occurs from country to country: retention of nesting grounds in Canada through government purchase of these grounds is of benefit to the United States (or at least to its hunters). But by the same token, U.S. government expenditures for wintering-ground refuges are of benefit to Canada. Again some balancing out occurs, this time from a country-to-country point of view.

Hunter-to-hunter externalities also may be identified: (1) The more hunters of a given number of birds, the fewer birds per hunter. (2) Birds winging their way to California, downed by Washington and Oregon hunters, mean fewer birds for Californians. We studied this problem in greater detail in a section of chapter 5. (3) If birds with a Pacific flyway orientation crowd out fowl from other flyways in the prairie breeding grounds, Pacific flyway hunters will benefit at the expense of others.[7]

[6] *Competition for Wetlands in the Midwest* (Baltimore: Johns Hopkins Press, for Resources for the Future, 1971), pp. 9–40.

[7] Orientation of waterfowl to wintering grounds is a fascinating unsolved biological problem. Waterfowl are thought to return in later years to the same wintering area that they first visited. But the mechanism determining the first year's wintering ground is unknown. Banding of duckings believed to be broodmates, for several species, yielded evidence of separation of broodmates in migration and wintering. (See R. K. Martinson and A. S. Hawkins, "Lack of Association Among Duck Broodmates During Migration

SUMMARY

Our model, culminated in the previous chapter, represents the first effort to join economic with physical relationships in order to determine the optimal number of ponds, breeders, and seasonal kill. The many inadequacies of the present model are best characterized in general terms. It is basically a time-dependent analysis of a single plant producing one good with two factors of production. The good is purchased in a single market. There is no uncertainty (with the exception of simulation experiments in chapter 4). The real world, and a more adequate model, calls for a multistage, spatially differentiated, stochastic production process yielding many goods from many quality differentiated inputs. The goods are sold in many spatially differentiated markets.

Much of the needed additional research is in the province of waterfowl biologists, and much of the research should be done in conjunction with economists. We need to know more about what determines routes and destinations of the fall flights of many, if not all, of the species of dabblers, divers, and geese. We need basic production data for species other than mallards, in addition to careful, independent identification of crucial rates of natural mortality. We require additional studies on crippling loss fractions. We must learn more about species interaction and competition for available breeding grounds and wintering habitat. And how alike are hunter valuations across flyways and through time?

Despite all the limitations of our present model, we confess considerable belief in its results. Although we would not claim that the marginal value of a bagged waterfowl is \$3.10, or \$3.29, we would be very surprised if, in time, a more precise model were to produce a value for ducks much greater than this order of magnitude.[8] Further, considering that pond costs as high as \$17 (on an annual basis) signal an increase in the number of ponds from recent annual counts, we would be most surprised if a more precise model were to indicate that existing permanent wetlands should be drained for agricultural use.

In a broad sense, we have attempted to identify in both theoretical and

---

and Wintering," *Auk*, vol. 85, October 1968, pp. 684–686.) The number of broods involved in the study is small, but the evidence is not supportive of an instinctual mechanism, nor of leadership by the parent hen.

[8] A note of caution to the reader is sounded: one cannot attribute all of the value of a waterfowl to the breeding grounds, as we have done under the assumption that wintering grounds are adequate, and then apply the same waterfowl value figure to justify purchases of resting, wintering, and hunting grounds. To do so would count benefits several times over.

empirical terms the costs and benefits derived from hunting waterfowl. The value of such analyses in decision making has been summed up by Kenneth E. Boulding:

Even though economic measurement may be abused, its effect on the formation of moral judgments is great, and on the whole I believe beneficial. The whole idea of cost–benefit analysis, for instance, in terms of monetary units, . . . is of enormous importance in the evaluation of social choices and even of social institutions. We can grant, of course, that the . . . dollar . . . is a dangerously imperfect measure of the quality of human life and human values. Nevertheless, it is a useful first approximation, and in these matters of evaluation of difficult choices it is extremely useful to have some first approximation that we can then modify. Without this, indeed, all evaluation is random selection by wild hunches. It is true, of course, that cost–benefit analysis of all sorts of things, whether of water projects, other pork barrel items, or in more recent years weapon systems, can be manipulated to meet the previous prejudices of people who are trying to influence the decisions. Nevertheless, the fundamental principle that we should count all costs, whether easily countable or not, and evaluate all rewards, however hard they are to evaluate, is one which emerges squarely out of economics and which is at least a preliminary guideline in the formation of the moral judgment, in what might be called the "economic ethic."[9]

[9] "Economics as a Moral Science," *American Economic Review*, vol. 59 (March 1969), pp. 7–8.

Appendix

TRANSMITTAL LETTERS AND QUESTIONNAIRE

UNIVERSITY OF WASHINGTON
Seattle, Washington 98105

Department of Economics

Dear Waterfowl Hunter:

The University of Washington is engaged in a study of waterfowl and waterfowl hunting in the Western States. We hope the results of the study will help to improve the management of waterfowl resources. Through a random selection process you have been chosen to receive our questionnaire.

Only you, the waterfowl hunter, can give us the answers we need. Will you please help us by filling out the enclosed questionnaire and returning it in the self-addressed envelope which requires no postage? We have tried to make the questionnaire as short as possible. Although some of the questions may be difficult for you to answer accurately, please try to give your best guess.

You have our word that your individual answers will be held in strict confidence and will be used only for the purposes of this study. Your response will be combined with many others and only group totals will be used in the study results. The number that appears on your questionnaire allows us to remove your name from our list when the questionnaire is returned, so that we may avoid bothering you again.

If you have any comments you would like to make about the questionnaire, feel free to do so on the back of this letter and return it along with the questionnaire.

Thank you for your cooperation.

Best wishes for good hunting,

Dr. Gardner Brown, Jr.
Assistant Professor of Economics,
Institute for Economic Research

UNIVERSITY OF WASHINGTON
Seattle, Washington 98105

Department of Economics

Dear Waterfowl Hunter:

We seem not to have received back the waterfowl hunting questionnaire sent to you a few weeks ago. We enclose another copy in case something happened to the first one. Won't you take a few minutes to fill it out and return it to us in the self-addressed postage-paid envelope? *Your* answers *are* important.

We emphasize that your individual answers will be held in strict confidence and will be used only for the purposes of this study. Your response will be combined with many others and only group totals will be used in the study results. The number that appears on your questionnaire allows us to remove your name from our list when the questionnaire is returned, so that we may avoid bothering you again. If you have any comments you would like to make, feel free to do so on the back of this letter and return it along with the questionnaire.

If you have already mailed the original copy back to us, we are sorry to have bothered you again.

Thank you once more for your help.

Best wishes for good hunting,

Dr. Gardner Brown, Jr.
Assistant Professor of Economics
Institute for Economic Research

WATERFOWL HUNTING QUESTIONNAIRE

We use the term "waterfowl hunting season" or just "hunting season" often in our questions. By this we mean the open waterfowl hunting period usually starting in October of one year and ending in many areas in January of the following year. To emphasize that a single season may overlap two calendar years we refer, for instance, to the "1968–69 hunting season."

Question 1. During approximately how many seasons have you hunted waterfowl, *including* the 1968–69 season?
_______________ seasons

Question 2. Did you purchase a Migratory Bird Hunting Stamp ("Duck Stamp") for the 1968–69 season? Please indicate your answer by a check mark below.
_______ Yes _______ No

IF YES, please go on to Question 3.

IF NO, please check the reason why you did not purchase a stamp. (Please check *only* the major reason.) Then skip directly to *Question 9a* and answer the remaining questions.

_______ Waterfowl hunting too expensive
_______ Not enough time available to hunt
_______ Too far to drive to hunting area
_______ Insufficient number of waterfowl
_______ Areas of waterfowl hunting too crowded with other people
_______ Other reason (please explain) _______________

Question 3. Kindly indicate below the kinds of waterfowl you prefer to hunt. Use a "1" for your *first* choice, a "2" for your *second* choice (if any), etc. Please do not go beyond your third choice.

Ducks:
_______ American Widgeon
_______ Mallard
_______ Pintail
_______ Shoveler
_______ Teal
_______ Goldeneye, Redhead or Scaup

_______ Other varieties of ducks not listed at left
_______ Ducks in general, don't particularly care what variety
_______ Geese
_______ Coots

Question 4. Approximately how many waterfowl did you shoot and retrieve during the 1968–69 season? (If the answer is zero, please enter "0.")

_______________ waterfowl

Question 5. On about how many days did you hunt waterfowl during the 1968–69 season? (If the answer is zero, please enter "0.")

_______________ days

Next, we would like to ask you a question involving an entirely fictitious situation; it is the only way we can obtain the desired information. The question may seem difficult and may take some thought, but we would like your best guess.

Question 6. Suppose you have the right to hunt waterfowl just as you have had that right during the last season and the seasons before it. But also suppose that you could sell your right to hunt waterfowl for *a season*, and if you did sell the right, you yourself could not hunt waterfowl during that season. You would set your own price and the choice would be entirely up to you whether or not you sold this right.

We emphasize that this situation is entirely fictitious—no one is going to restrict waterfowl hunting on the basis of this questionnaire and no one could actually buy or sell this natural right.

But, WHAT IS THE *SMALLEST* AMOUNT YOU THINK YOU WOULD TAKE TO GIVE UP YOUR RIGHT TO HUNT WATERFOWL FOR A SEASON—SAY, 1968–69? Please check the most appropriate answer below.

_______ $0.00 to $2.49
_______ $2.50 to $4.99
_______ $5.00 to $9.99
_______ $10.00 to $19.99
_______ $20.00 to $29.99
_______ $30.00 to $49.99
_______ $50.00 to $74.99
_______ $75.00 to $99.99
_______ $100 to $199
_______ $200 to $299
_______ $300 to $399
_______ $400 to $499
_______ $500 to $749
_______ $750 to $1000
_______ Over $1000 (please specify amount $_______________)

Question 7. About how much do you figure your total waterfowl hunting costs were for the 1968–69 season? (An obvious cost would be shotgun shells. We are interested in what *you* consider your costs to be. We therefore prefer not specifying cost categories, thus leaving the matter up to you.) After you have given the question a little thought, please check the answer which you feel best represents your *total* costs *for the season.*

______ \$0.00 to \$2.49	______ \$100 to \$199
______ \$2.50 to \$4.99	______ \$200 to \$299
______ \$5.00 to \$9.99	______ \$300 to \$499
______ \$10.00 to \$19.99	______ \$500 to \$749
______ \$20.00 to \$34.99	______ \$750 to \$999
______ \$35.00 to \$49.99	______ \$1000 to \$1499
______ \$50.00 to \$74.99	______ \$1500 to \$1999
______ \$75.00 to \$99.99	______ \$2000 or more (please specify amount \$______)

Since we have been talking about costs, we would now like to ask you another question on the same subject, but this one again involves an entirely fictitious situation. Again, the question may seem difficult and take some thought, but we would like your best guess.

Question 8. Suppose that your waterfowl hunting costs for the 1968–69 season were greater than you estimated in Question 7.
Assume these increased costs in no way affected general hunting conditions.
ABOUT *HOW MUCH GREATER* DO YOU THINK YOUR COSTS WOULD HAVE HAD TO HAVE BEEN BEFORE YOU WOULD HAVE DECIDED NOT TO HAVE GONE HUNTING *AT ALL* DURING THAT SEASON?

We emphasize that the dollar amounts given below are intended to represent imaginary increased costs, that is, costs over and above the actual costs you estimated in Question 7. Please check the answer below that you consider most appropriate.

______ \$0.00 to \$2.49	______ \$100 to \$199
______ \$2.50 to \$4.99	______ \$200 to \$299
______ \$5.00 to \$9.99	______ \$300 to \$399
______ \$10.00 to \$19.99	______ \$400 to \$499
______ \$20.00 to \$29.99	______ \$500 to \$749
______ \$30.00 to \$49.99	______ \$750 to \$1000
______ \$50.00 to \$74.99	______ Over \$1000
______ \$75.00 to \$99.99	(please specify amount \$______)

Question 9a. Please indicate your approximate age:

______ Under 16 years	______ 35–44 years	______ 55–64 years
______ 16–24 years	______ 45–54 years	______ 65 years or more
______ 25–34 years		

Question 9b. Kindly indicate the category that represents your highest level of formal education:

______ 8th grade or under (elementary school)
______ 9th through 12th grade (high school)
______ Beyond 12th grade (college)

Question 10. The number of persons in my household, including myself, is

______ 1	______ 4	______ 7
______ 2	______ 5	______ Over 7 persons
______ 3	______ 6	(please specify number ______)

Question 11. The annual income after taxes in my household is about

______ Under \$2500 ______ \$10,000 to \$14,999
______ \$2500 to \$4999 ______ \$15,000 to \$19,999
______ \$5000 to \$7499 ______ \$20,000 to \$29,999
______ \$7500 to \$9999 ______ \$30,000 or more (please specify to nearest \$10,000 \$______)

Question 12. During the 1968–69 season did you at any time hunt for waterfowl on private property or a private club where you, yourself, had to pay in some way for the right to hunt?

______ Yes ______ No

IF YES, will you please give your estimates on the items below?

The number of waterfowl I shot and retrieved on private "pay land" during the 1968–69 season was about ______ waterfowl. (If zero, please enter "0.")

The number of days I hunted on private "pay land" during the 1968–69 season was about ______ days.

My *total* costs for hunting on private "pay land" during the 1968–69 season were about \$______.

COMMENTS OF QUESTIONNAIRE RESPONDENTS

In the letters of transmittal, hunters were invited to make such comments as they desired. The ratio of favorable comments to those unfavorable was approximately twenty to one. However, as unfavorable comments were often more colorful, no attempt is made to maintain that ratio in the selected direct quotes below. If a number appears in parentheses after a comment, it indicates the number of people who made similar remarks.

This is the story of my husband's life.
The poor duck hunter in his blind
He's cold in front and wet behind
It's been eight hours since he's fed
And twenty since he's been in bed
He's paid a hundred bucks or more
And shot until his arm is sore
But now the ducks have gone away
Yet he has hopes for a better day
If his Christmas message is behind
It's cause he's still there in his blind.

. . . [The] set of questions were not filled out because we thought the questions were either so stupid or so personal that we could see no reason for filling them out.

If you need funds to help with this research, let me know and I will try to take up a collection with my friends.

Your question is asinine. It is about time you stop wasteing the sportsmans money on such trash. Go try to make your self an honest living for a change.

To give you an example of how I feel about duck hunting I even gave up smoking because the nicotine collected on the reed of my duck call causing it to have bad notes.

Studies like these are of great importance. If I can help please let me know.

I would appreciate any further information you could send me in regards to this study since I am quite interested in anything that might improve or aid in the management of waterfowl.

I would hope that we don't pay taxes to find out information of this irrelevant type.

An honor and a pleasure to be of service.

If you want information on hunter success ratios, write to the Department of Fish and Game.

How wonderful to have research done on a subject that matters!

I wouldn't sell my right to hunt at any price! . . . it is for rights like these that our forefathers fought and died.

I did not hunt this year. I've lost interest in killing any animals. [age 17]

If they had any sense they'd close the season for a year. (2)

Worst season ever. (33)

Please send the results of this study. (31)

Pure Communist.

I would like to know why they are draining the swamp land, taking away the nesting ground from the waterfowl.

Extend the season later in the year. (24)

Please forgive my tardiness, but I wanted to show [the questionnaire] to all my goose hunting friends, and believe me there are a *lot* of them!

My husband drowned while duck hunting, but I have tried to fill out the questionnaire for him.

I belong to Ducks Unlimited and help the Fish and Game Department each year with surveys. Furthermore, for every bird I shoot I buy two from a hatchery and release them in the field.

In regard to question 6 isn't this somewhat like how much would you sell your citizenship for?

I wouldn't sell my hunting rights, and I wouldn't sell my wife either.

If I sold my right to hunt ducks, I would hunt pheasants instead.

. . . in my younger days, I was an avid sportsman. Occasionally, I will purchase a hunting and fishing license and all stamps with no intent of use. I do this to contribute a small share, monetarily, to the agencies involved and in thanks for the past years I did enjoy.

Duck stamp money should be used near the area where it was raised.

Put pictures of protected species on the duck stamps so more hunters would know what they look like and not shoot them.

I didn't hunt. I feel I should leave the game and birds for the kids who come after us.

My son is in the Army. He could tell you how many Viet Cong he killed, but not waterfowl.

How about *you* sending us a questionnaire filled in by *yourself?*

The hunters increased and the ducks decreased. Now I would rather see the birds than kill them.

Glad to be of service.

. . . this form is less than useless, frivolous, and basically has nothing to do with my hunting activities.

I would also like to express my interest in what you are doing. . . . I would appreciate your sending me some facts on your study!

Keep up the grate work.

Obviously, these waterfowl questions are a phony front for your other questions.

I will *never* hunt on private pay land. (18)

It's those duck clubs that ruin the hunting for everyone else. (7)

I do not know you, what your project is, or who, or what, is paying for the costs involved. I do know that information is not mandatory by law regardless of its nature and is no one's business but my own.

Enclosed is a list of my fellow hunters who would be equally eager to help in any way possible.

Stricter enforcement of regulations. . . . (5)

Hughes doesn't have enough. [response to question 6]

Idiots in dune buggies should be banned during hunting seasons.

No wonder with money being spent on nonsense like this there's no money to buy refuges.

Glad to help in any way possible.

Alternate hunting years to alleviate crowded conditions.

I find both [questions] #6 and #8 to be hypersensual and asinine.

I am 70, and my hunting days are nearly over. I look back on my years of hunting with great pleasure, but I also recall how severely the fall flights have decreased and I worry whether those who come after me will have any birds to hunt at all.

I'm only 13, but I didn't buy a duck stamp, but I answered this anyway because I went hunting. Last year I bought a stamp because the sport shop salesman said I needed one, but I didn't. It was fun answering this!

given quality for a period, but not with the possibility of doing without alternative activities, such as hunting upland birds, which have similar subsidiary satisfactions associated with them. Hence the reported net benefit may be attributed to the waterfowl bagged.[25] The Davis experiment had recreationists report total net benefits, and we prefer not to tamper with this formulation directly. Instead, marginal valuation is obtained through a first derivative of a function having total net benefits as the dependent variable.

The remainder of this chapter will be devoted to a discussion of why, and in what manner, bagged waterfowl may be expected to have a positive marginal value. The next chapter will report the empirical procedure and the results of our attempts at measuring the value of bagged waterfowl to hunters.

MARGINAL VALUATION OF WATERFOWL

For ease of exposition, the following assumptions are made: (1) the hunter does not distinguish among waterfowl, i.e., the value of a bird is not influenced by its species, age, or sex; (2) the kill probabilities are constant for each hunter on each hunting site during the season, and each hunter is aware of the probabilities; (3) each hunter hunts waterfowl at only one site during the season; (4) all relevant costs are known to each hunter at the beginning of the season.

From these assumptions it follows that a representative hunter will consider any waterfowl shot by him to be equal in value to any other he might have shot. At the start of the season, he will choose the hunting ground that will maximize his total net welfare, knowing in advance the expected kill he will obtain at each site. Since all costs at the chosen site are known in advance, this will be interpreted to mean that all costs to the hunter may be regarded as variable and therefore reflected in marginal cost curves. Furthermore, the number of days the individual will hunt is known to him at the beginning of the season. Nothing in the assumptions and their implications prohibits: (1) the kill probabilities varying from site to site; or (2) the daily kill on an individual site varying from hunter to hunter.

Waterfowl hunters are subject to two institutional limits: the maximum

[25] An exception would occur if the various satisfactions of the recreation day are synergistic, i.e., if the satisfaction from the other activities and experiences of the day are greater when they are combined with waterfowl hunting than when hunting is missing. In such cases, the "synergistic benefit" unavoidably would be included in the measured net benefit.

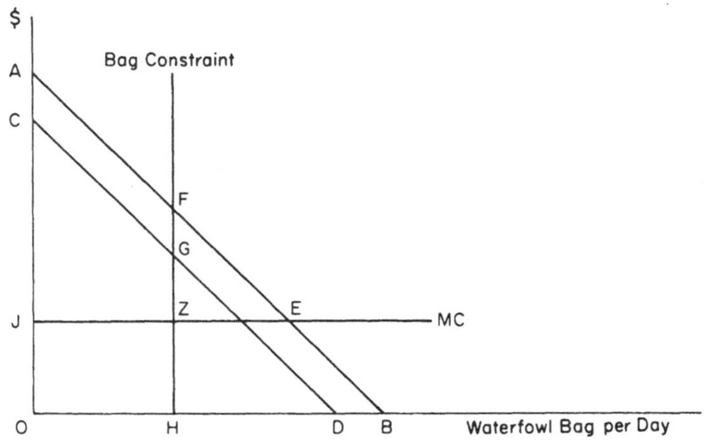

Figure 4. A hunter's demand and cost curves for bagged waterfowl.

number of waterfowl that may be bagged in one day and the maximum number of days during which waterfowl may be shot.[26] In the United States these bag and day limits are primarily determined by the federal government, which offers each state several combinations of bag and day limits designed to give approximately the same total kill for the season. Each state then chooses the option it prefers.

Line *AB* in figure 4 shows the income-compensated demand curve for bagged waterfowl for a representative hunter on his first day of hunting during the season.[27] The downward-sloping curve implies that each waterfowl bagged gives the hunter less satisfaction, measured in dollars, than the bird previously shot. The marginal cost of each bird is assumed constant. Since, as stated, all costs are considered variable, any costs that more conventionally would be recognized as seasonal or daily are apportioned to the marginal cost curve for bagged waterfowl. This artifice will allow a heuristic mapping between graphs.

Were the hunter subject to no constraints, he would shoot birds until he reached point *E* in figure 4, if that many birds were available on the chosen site. In such case, the gross benefits from the last bird shot would be equal to cost, and the marginal bagged waterfowl would yield no net

[26] There is a third constraint, the maximum number of birds a hunter may have in his possession. The number varies from state to state but is usually either the bag limit or twice the bag limit. The possession limit here is considered relatively minor and is neglected in the analysis.

[27] For simplicity, all demand curves in figures 4 and 5 are depicted as linear continuous functions.

benefit. This is consistent with economic theory regarding conventional goods, where the consumer will continue to buy a good until the marginal unit yields gross benefits just equal to the cost of that unit. However, waterfowl are not conventional goods, and the hunter is assumed to be unable to reach *E* due to the constraint imposed by the vertical line at *H*, representing either the legal bag limit or the number of birds that may be shot on the hunting ground, whichever is the lesser.

The relevant constraint may be thought of as a cost constraint that becomes very large at some point: a hunter in Utah on land with a probable yield of two waterfowl per day may travel to California to obtain six waterfowl per day, and so on. In addition, the cost of exceeding the bag limit may be both a fine and social stigma. For a first approximation, we may consider such cost infinite. Viewing the constraint as one of cost, the hunter would proceed to maximize his net benefit by equating his marginal gross benefit with marginal cost through shooting waterfowl each day up to his constraint. Thus, on each hunting day, the hunter reaches a constrained equilibrium. This should not be interpreted, however, as implying that an additional waterfowl shot per day would have no positive net value. It well may have, and the value would be represented by the amount the individual would pay for the relaxation of the constraint by one waterfowl per day. On the first day of hunting, the hunter's costs are represented by the area *OJZH*, and his net benefit, or consumer's surplus, equals the area *JAFZ*.

Alternatively, as in figure 5, the net benefits and costs may be attributed to the recreation day, where *KL* represents the income-compensated

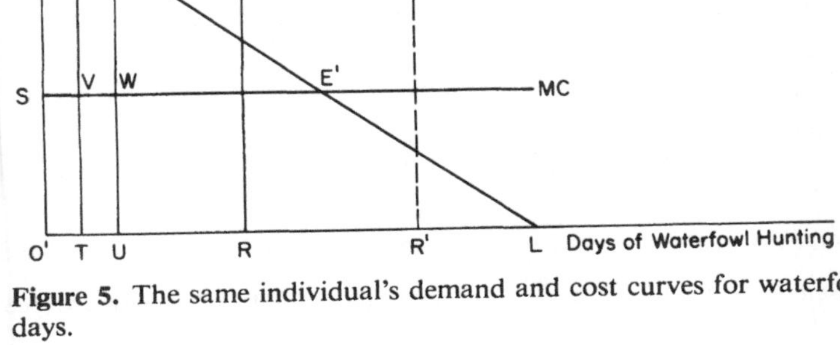

Figure 5. The same individual's demand and cost curves for waterfowl hunting days.

demand curve of the hunter for hunting days during the season, and *O′T* represents the first day of hunting. *O′SVT* represents the cost of the day's hunting and equals and corresponds to *JAFZ* in figure 4.[28]

If the hunter faces no constraint, he will hunt the number of days required to reach equilibrium *E′*. However, a constraint related to the season's day limit may or may not operate. The number of days an individual hunter will engage in a recreational activity is a function of the day limit and his entire preference field for the use of his time. Hence his hunting days will be determined either by the constraint or by his having reached the unconstrained equilibrium, whichever occurs first. The constraint may also be considered a cost constraint, where the marginal cost of an additional day suddenly becomes infinite, e.g., if the individual spends one more day hunting he will lose his job or intolerable marital conflict will ensue. In figure 5, the constraint would be operative at *R* and irrelevant at *R′*.

The second day of hunting is represented in figure 5 by *TU*. Since the demand curve is downward-sloping and the marginal cost per day is assumed constant, the net benefits must be less on the second day than on the first day. Hence, in figure 4 the demand curve for bagged waterfowl on the second day, *CD*, must lie below the demand curve for the first day, *AB*. The net benefits from the second day of hunting, represented as *JCGZ* in figure 4, correspond to *VMNW* in figure 5. Although only two demand curves are shown in figure 4, a curve exists for each day of hunting and each curve is lower than the preceding one.[29]

Let us assume that the constraints in each graph are operative for the representative hunter and then observe the results of relaxing one limitation while holding the other constant.[30] If the bag limit is increased (or more waterfowl are available) while the season length remains the same, the constraint in figure 4 shifts to the right and (assuming that curves in figure 4 do not shift as a result) the demand curve in figure 5 shifts outward, as each recreation day brings greater net benefits than it did before. This effect may be thought of as a quality change; the quality of the recreation day has benefited from the permitted increased number of waterfowl bagged per day.

[28] It should be noted that in our terminology the net benefits from the day's hunting do not include the subsidiary benefits of the recreation experiences mentioned earlier in the text, and therefore the net benefits properly may be attributed to the waterfowl bagged.

[29] In theory, the curves could intersect. To simplify, it is assumed arbitrarily that they do not.

[30] The assumption of independence is, of course, artificial. It does, however, produce a situation that can be readily analyzed.

Alternatively, if the authorities increase the season length while holding bag limits constant, it is assumed that this will shift the constraint in figure 5 to the right. The hunter will hunt more days and additional demand curves on figure 4 will become relevant. We can consider this a quantity change; an increase in the permissible number of hunting days results in a greater number of waterfowl bagged.

Since many hunters are subject to one or both constraints, the net benefits to them of the marginal waterfowl will be positive. A peculiar situation exists in that a single item, an additional waterfowl, can affect each of two margins: a quality margin and a quantity margin. The bird may be taken either as an additional bird shot on one of the (constant number of) hunting days, or as a benefit resulting from shooting (a fraction of) an additional day, with kill per day constant. Which margin is affected depends upon which constraint has been relaxed.

Furthermore, it is entirely possible that the two margins diverge: one constraint may be more binding than the other. As an example, consider the situation of the individual represented by figures 6 and 7. By shooting the bag limit *ON* (figure 6) on each of two days of hunting, he reaches unconstrained equilibrium *E′* (figure 7) in terms of days, given the bag limit. He would gain no positive net value from an additional day's hunting, and therefore would gain no positive net value from an additional waterfowl shot on a fraction of an additional day. On the other hand, the

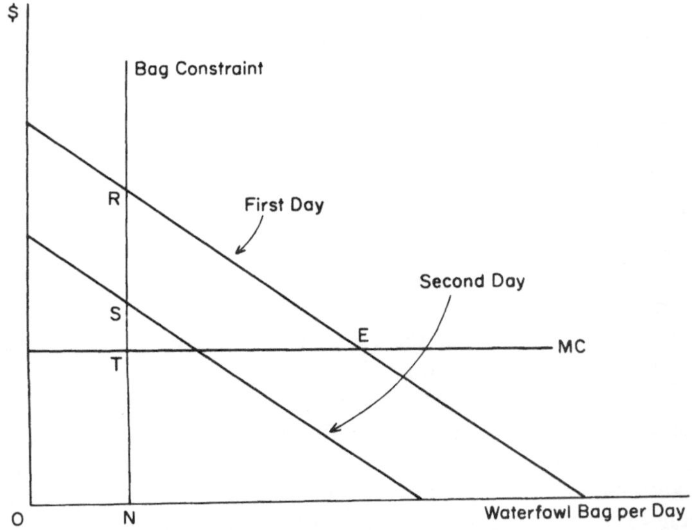

Figure 6. Illustration of active constraint on a hunter's daily bag.

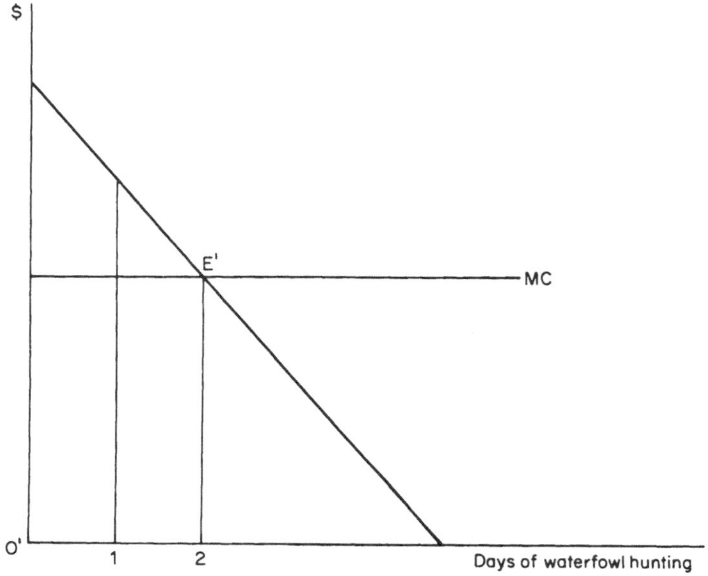

Figure 7. Illustration of unconstrained equilibrium of hunting days consistent with figure 6.

marginal value of an additional waterfowl per hunting day is positive, represented in figure 6 by line length *RT* for the first day and *ST* for the second. Hence, an additional bird taken as a probabilistic fraction of an additional waterfowl per day has a positive net value.

A MATHEMATICAL FORMULATION

The model also can be described through a mathematical equation for an individual hunter:

$$V = V(Y, U, D/Z, Z) \qquad (2\text{-}2)$$

where V is a measure of consumer's surplus; Y is a measure of income, preferably discretionary income; U represents some measure or measures of taste; D is one hunter's bagged waterfowl for the season; and Z is the number of days the individual hunted during the season.

We impose certain constraints on the functional form of equation (2-2): namely, that the first partial derivative of V with respect to each independent variable is positive, and the second partial derivative with respect to each variable is negative. Downward-sloping demand requires the imposition on the last two variables; the remaining constraints are by supposition.

The first partial derivative of V with respect to (D/Z), denoted by $V_{(D/Z)}$, is an approximation of the marginal value of an additional waterfowl taken on *each* day the specified individual hunts during the season.[31] To obtain the approximation of the marginal value of one additional waterfowl, the partial must be divided by Z. Hence

$$\frac{V_{(D/Z)}}{Z}$$

represents the "quality margin"—the quality of the recreation day has been increased by shooting a probabilistic fraction of an additional waterfowl per day, holding constant the number of days hunted.

By similar reasoning,

$$\frac{V_Z}{D/Z}$$

approximates the "quantity margin"—the additional waterfowl has been taken by shooting an additional fraction of a day, holding constant the number of waterfowl killed on each hunting day.

As noted in the preceding section, there is no *a priori* reason why the two marginal valuations will be equal for any given hunter or for a number of hunters taken as a group. Which constraint is the more binding depends upon the functional forms of the demand curves in figures 4 and 5. Relaxing the assumption that all costs are variable, thus allowing a divergence in figures 4 and 5 between the variable costs of a day's kill and the variable costs of a hunting day, provides an additional reason for a difference in the two marginal valuations.

[31] From equation (2-2),

$$\frac{dV}{d(D/Z)} = V_Y \frac{dY}{d(D/Z)} + V_U \frac{dU}{d(D/Z)} + V_{(D/Z)} + V_Z \frac{dZ}{d(D/Z)}$$

Assuming

$$\frac{dY}{d(D/Z)} = 0, \frac{dU}{d(D/Z)} = 0, \frac{dZ}{d(D/Z)} = 0$$

then

$$\frac{dV}{d(D/Z)} = V_{(D/Z)}$$

Setting the above terms equal to zero requires no explanation, except for the last term. That

$$\frac{dZ}{d(D/Z)} = 0$$

is consistent with the assumptions in the main text. But, of course, Z may in fact change as a result of a change in (D/Z). Hence, the partial derivative $V_{(D/Z)}$ is an approximation of the full derivative, since it assumes that the above change does not occur.

chapter three Valuation of Waterfowl: An Empirical Test

THE HUNTER SAMPLE

In order to apply equation (2-2) to several individuals, data are required for consumer's surplus, income, tastes, waterfowl bagged per day, and days of hunting.

Available resources restricted this study to the waterfowl hunters in the seven western states that lie wholly within the boundaries of the Pacific flyway: namely, Arizona, California, Idaho, Nevada, Oregon, Utah, and Washington. The necessary data were obtained from the hunters through questionnaires sent to a sample of 4,900 hunters randomly selected from a list provided by the U.S. Bureau of Sport Fisheries and Wildlife. These individuals purchased legally required federal waterfowl hunting stamps during the 1967–68 season, and many of them responded to the Bureau's waterfowl kill survey for the same year. Although the names and addresses of hunting license purchasers presumably could have been obtained from each state and the questionnaire sent to a sample of these hunters, many of those contacted would not have hunted waterfowl and a waste of money and energy would have resulted.

THE QUESTIONNAIRE FORMAT

A waterfowl hunting questionnaire pertaining to 1968–69 was developed that inquired directly or indirectly about the number of seasons of hunting, the purchase of a waterfowl hunting stamp, waterfowl species preference, waterfowl bagged during the season and number of days of hunting, costs, consumer's surplus, age, education, number in family, income after taxes, and hunting on private land. Letters of transmittal, the questionnaire, and selected responses are reproduced in the appendix at the end of this book.

In the belief that hunters would be more likely to respond to a brief questionnaire, questions were carefully restricted and most were designed to be answered with no more than a number or a check mark. Obviously an information cost is incurred in minimizing the number of questions. For instance, a breakdown of costs attributable to killing a bird, to a hunting day, to the season as a whole, and to more than one season would have been of interest. But because these questions would have increased the length of the questionnaire considerably and this might have decreased the number of responses, they were not asked. The cost question asked was somewhat novel: seeking information on the hunter's own perception of his costs implies that hunters make decisions based on what they consider their costs to be, rather than on what an economist believes should be included or excluded.

The first page of the questionnaire was designed to put the respondents at ease through inquiries on expected subjects—years of hunting experience and preferences regarding species of waterfowl, for example. The sensitive areas of income and educational level were left to the fourth page, while the inner pages contained inquiries regarding costs and the crucial measures of consumer's surplus.

In the middle of February 1969, shortly after the close of the hunting season, printed and numbered copies of the questionnaire were mailed to the hunters in the sample. In the first week of April, another copy was mailed to those who had not responded.

CONSISTENCY OF THEORETICAL ASSUMPTIONS WITH EMPIRICAL ANALYSIS

In chapter 2, a number of simplifying assumptions were made for ease in presenting our theory. Their consistency with the empirical analysis requires discussion.

The assumption that, to the hunter, all waterfowl are the same, permits all retrieved waterfowl to be reported as a single item on the questionnaire. Separate estimates by hunters of dabbling duck kill, diving duck kill, and goose kill might be subject to considerable memory bias. Furthermore, this assumption allows a single marginal valuation.

In the initial theoretical presentation, all costs for the season are treated as variable costs. However, this treatment was primarily for expository purposes and was relaxed later.

The theoretical formulation also assumes that each hunter shoots at one site for which the kill probabilities are known and constant for that hunter throughout the season. A single pair of graphs with stable constraints (figures 4 and 5) could then be used to illustrate the situation faced by the representative hunter. But these assumptions, particularly that of hunting at a single site, are untenable in reality. In the key Pacific flyway state of California, for instance, the number of hunting days per hunter is restricted on state and federal refuges where the kill probabilities are excellent. When the restriction becomes binding, hunters go to other sites. The assumptions are retained in the empirical analysis, but in specialized form. It is assumed that data from each hunter's questionnaire response reflect a single (composite) hunting ground, and incorporate weighted averages of the differing characteristics of the sites actually hunted, and weighted averages of the kill probabilities if they changed throughout the season on any site.

QUESTIONNAIRE RESULTS

The questionnaire produced 2,455 usable responses, for a response rate of 50 percent. The first mailing resulted in 1,792 responses and the second in 663.

It is particularly important to check the validity of equation (2-2), namely, $V = V(Y, U, D/Z, Z)$, for a group of individuals through regression of data obtained from the questionnaire responses. The regression results can be compared with those from variations of (2-2), and the marginal value of a waterfowl to the hunter can be quantified using the parameter estimates from the regressions.

Choice of the Consumer's Surplus Measure

The questionnaire attempted to elicit an estimate of both the willingness-to-sell and the willingness-to-pay variations of consumer's surplus from each respondent who hunted in 1968. The question dealing with the willingness to sell (through a hypothetical sale of hunting rights for a season) provoked a highly emotional response in many individuals, as they objected to even the thought of selling their right. The percentage of hunters who wrote that they would not sell their right at any price was 12.4, and the percentage who named a very high figure such as one million dollars was 1.4 percent. Eliminating these responses, the mean value was $1,044.

The willingness-to-pay question elicited no discernible emotional response in the large majority of respondents. The question, which followed one on total waterfowl hunting costs for the season, asked what minimum amount costs would have had to have risen before the respondent would have been discouraged from hunting during the season, assuming general hunting conditions remained unchanged. Many individuals who said they would not sell their right to hunt at any price indicated, without comment, relatively low and seemingly quite rational dollar amounts in reply to the increased cost question; the mean amount was $247.

The valuation under willingness to pay is the measure of primary interest in this study and was used as V in all regressions. Why is it the relevant measure? Because waterfowl require wetlands for breeding and for rearing young, and because the large majority of ponds and potholes threatened by drainage now and those most likely to be threatened in the future lie in the prairie region of North America that is private farmland. Given these property rights, the appropriate measure of consumer's surplus is the additional amount individuals would be willing to pay to preserve existing wetlands from destruction by man or to increase the amount of wetlands available through creation by man.[1]

The valuation under conditions of the compensating variation is the one of primary interest in this study and was used as the appropriate measure for V in all regressions. Since ponds and potholes that produce waterfowl may be legally drained for other uses, the relevant consumer's surplus measure is the one that reflects the additional amount individuals would be willing to pay to continue to hunt.

Measures of both valuations were sought for two reasons: (1) asking the willingness-to-sell question before the more important willingness-to-pay question allowed some prior psychological adjustment to a hypothetical question on the part of the respondents; (2) as stated in chapter 2, comparison of accurate measures of the two consumer's surpluses would provide information regarding the income elasticity of the good. (In this particular instance, the good is probably best defined as hunting days.) However, the dollar amount of the equivalent variation reported previously is felt to be emotionally biased, and no claim is made to even a first approximation of the true valuation. On the other hand, claim is made to having obtained a first approximation of the willingness-to-pay consumer's

[1] We argue that willingness to pay is appropriate here under conditions of both expansion and contraction of the ecological system. But such is not always the case, at least in the instance of contraction. On the assumption that people have a right to sport fish, perhaps in a prescriptive sense since legal rights are not well defined, the willingness-to-sell approach is suitable for decisions involving a prospective decrease in the sport fish catch. This assumption and a modification of the Davis method, the only known technique compatible with willingness to sell, were used by Stephen B. Mathews and Gardner M. Brown in *Economic Evaluation of the 1967 Sport Salmon Fisheries of Washington*, Technical Report No. 2 (Olympia: Washington Department of Fisheries, 1970).

surplus of the questionnaire respondents. (The terms consumer's surplus and net benefits henceforth will refer to this valuation only.)

Regression Results

Multiple linear regression analysis was performed using data from the questionnaire responses, and the most satisfactory equation form and fitting was

$$\ln V = \underset{(t\text{ values})}{1.44} + \underset{(8.7)}{0.466 \ln Y} + \underset{(4.5)}{0.168 \ln S} + \underset{(5.3)}{0.141 \ln E} + \underset{(7.0)}{0.308 \ln D/Z} + \underset{(12.5)}{0.480 \ln Z} \quad (3\text{-}1)$$

$$\bar{R}^2 = 0.222, \quad s^2 = 1.70, \quad n = 1511$$

where Y is household income (after taxes) in thousands, S is number of seasons of waterfowl hunting, and E is annual cost of hunting.[2] The other variables are as previously defined. Corrected R^2 is denoted by $\bar{R}^2$.

All variables are significant at the 0.001 level, and the significance of the last two supports the theoretical formulation. Ahtough a higher $\bar{R}^2$ would be desirable, the result obtained here is acceptable for a cross section study involving no aggregation. The signs and magnitudes of the regression parameters are encouraging, as they meet the restrictions on the functional form stated in chapter 2, namely, that the first partial derivatives of V with respect to the independent variables must be positive and that the second partials of the independent variables must be negative.

These restrictions ruled out an untransformed linear regression. Besides the logarithmic transformation reported above, reciprocal and semilogarithmic transformations were tested. The reciprocal form gave an exceedingly poor fit. The semilogarithmic fit was satisfactory and as far as discernible was neither superior nor inferior to the logarithmic fit, with one exception. The intercept of the semilogarithmic transformation was negative and of sufficiently great magnitude in absolute terms to give negative estimates of V for some hunters in the sample, thus implying that they should not have gone hunting at all. This does not mean the semilogarithmic form is totally unacceptable, as primary interest lies in the marginal valuation of a waterfowl estimated through the values of first partial derivatives; in semilogarithmic form the intercept does not directly influence the value of the first partials. However, since the logarithmic trans-

[2] Many of the questions called for a response in terms of checking a range of values, e.g., income within the range of $5,000 to $7,499. The means of the ranges were used in the regression analysis.

Table 1. Means of Untransformed Variables

Variable	Symbol	Sample size[a] 1,692	1,511	1,122
Consumer's surplus ($)	V	$247	$256	$257
Household income after taxes ($1,000)	Y	$ 11.9	$ 12.1	$ 12.2
Seasons hunted (number)	S	16.1	16.2	16.7
Hunter's cost per season ($)	E	$377	$322	$301
Bagged kill per day (number)	D/Z	2.19	2.46	
Days of hunting per person (number)	Z	9.7	10.5	
Seasonal bagged waterfowl kill (number)	D	21.4	23.9	25.0
Bagged kill in 1967 (number)	D_{67}			25.7

[a] Of the 1,692 active hunters who gave responses to consumer's surplus, costs, income, and number of seasons of hunting, 1,511 reported at least one kill in 1968, and 1,122 reported at least one kill in both 1967 and 1968.

formation gives positive estimates of consumer's surplus for all hunters, and since considerable study has been devoted to its properties, the logarithmic form was adopted.

The transformation requires that the values for all variables be greater than zero. Only those questionnaire respondents who gave nonzero responses to all relevant variables were included in the regression.[3]

Of the 2,455 questionnaire respondents, 2,013 potential hunters had purchased a waterfowl hunting stamp and of these, 1,861 were active hunters engaging in at least one day of sport. Of the active hunters, 1,692 also gave responses to consumer's surplus, costs, income, and number of seasons of hunting.

The successful hunters who shot and retrieved at least one bird and who gave responses to the other variables numbered 1,511. The sample used in equation (3-1) consists of the individuals in this last group. The means for selected variables for this and the preceding group are given in table 1.

Choice of Independent Variables

Some readers may pause at the positive role of waterfowl hunting cost (E) in the determination of consumer's surplus. We see no *a priori* reason why cost cannot be a proxy measure for taste.[4] Of two hunters alike in important aspects such as income, success, and days hunted, one may

[3] Because the values of costs and income were the midpoints of ranges, the values could not be zero, although they could be quite low. Since hunting seasons included 1968, the value could not be less than one.

[4] Accepting consumer's surplus as being positively related to annual expenditures for waterfowl hunting, in a cross section study, does not imply that consumer's surplus is positively related to the price of specified hunting opportunities, such as the price paid for the right to hunt for a season on a particular piece of private land.

make larger expenditures, purchase a richer set of characteristics, and value the hunting experience more.

The ideal measure of income for explanation of recreation values would be discretionary income, i.e., income remaining after purchase of essentials (however defined). The best available substitute, household income after taxes, was used instead. The division of household income by the number of persons in the family made no distinguishable difference in results and was rejected by invoking Occam's razor.

On first glance, multicollinearity between variables D/Z and Z might be thought high, as Z appears in both. That this may not be the case is perhaps best illustrated by example. A person's daily wage rate and the number of days he works per week may be considered largely independent of each other, yet the daily wage rate might be expressed as the weekly wage divided by the number of days worked. The weekly wage may be thought of as a composite of two independent variables, and seasonal kill may be thought of as a composite of daily retrieved kill (D/Z) and days hunted. In the logarithmic transformation used in equation (3-1), the correlation coefficient between D/Z and Z was -0.15.

A number of sets of dummy variables were tested as additions to equation (3-1). Each set was added separately. The chosen dummy variable form affected the value of the intercept (of the logarithmic transformation). Dummy variables rejected as not having significance were those representing level of education, the fact of having paid to hunt on private land, state of residence, and response to first mailing.[6] That the dummy variable representing mailing response was not significant (t value = 0.57) is of interest. Mean values of the independent variables differed considerably in magnitude from the first mailing response to the second. For instance, the mean number of waterfowl was 4.8 birds greater for first-mailing respondents than for their second-mailing counterparts, and the mean consumer's surplus was greater by $54. That the dummy variable was not significant lends some support to the hypothesis that the regression parameters are valid over a broad range of hunters, perhaps including those who did not respond to the questionnaire.

[6] The significance was based on examination of t values. All dummy variables could be rejected at the 0.10 level except the set representing the seven states, where one pairing was significant at the 0.05 level. It is recognized that the t test is inappropriate where multiple comparisions are made, as is done with the states. The use of the t test in such cases tends to bias toward acceptance of the significance of some pairing. Considering the bias and the exceedingly high significance of the other variables, the dummy variables related to states were rejected.

Substitution of Seasonal Waterfowl Kill as an Independent Variable

Seasonal retrieved waterfowl kill, D, was added to the regression equation and retrieved kill per day, D/Z, and hunting days, Z, were removed. As explained previously, the added variable represents a composite of the two excluded.

The regression results were:

$$\ln V = 1.54 + 0.443 \ln Y + 0.163 \ln S \quad (3\text{-}2)$$
$$(t \text{ values}) \qquad (8.4) \qquad (4.4)$$
$$+ 0.149 \ln E + 0.409 \ln D$$
$$(5.6) \qquad (12.9)$$

$$\bar{R}^2 = 0.217, \quad s^2 = 1.71, \quad n = 1511$$

For purposes of comparison, regression equation (3-1) is repeated:

$$\ln V = 1.44 + 0.466 \ln Y + 0.168 \ln S \quad (3\text{-}1)$$
$$(t \text{ values}) \qquad (8.7) \qquad (4.5)$$
$$+ 0.141 \ln E + 0.308 \ln D/Z + 0.480 \ln Z$$
$$(5.3) \qquad (7.0) \qquad (12.5)$$

$$\bar{R}^2 = 0.222, \quad s^2 = 1.70, \quad n = 1511$$

Obviously, the substitution causes no untoward change in the magnitude of the regression parameters assigned to the independent variables common to both equations. The marginal waterfowl valuations implicit in equations (3-2) and (3-1) are now compared.

Comparisons of Marginal Valuations

The log-linear regression indicates a multiplicative relationship among the independent variables in their untransformed state. For instance, the untransformed counterpart of equation (3-2) is

$$V = q_0 Y^{q_1} S^{q_2} E^{q_3} D^{q_4} \quad (3\text{-}3)$$

or, substituting the parameter estimates,

$$V = e^{1.54} Y^{0.443} S^{0.163} E^{0.149} D^{0.409} \quad (3\text{-}4)$$

Thus, the marginal value of a waterfowl is the value of the partial derivative of V with respect to D,[6]

[6] Comparison of equations (3-3) and (3-4) shows that q_0 is assumed equal to $e^{1.54}$ where 1.54 is the value of the intercept in the logarithmic transformation regression (3-2). This means that a conditional median function of the variables in their untrans-

Table 2. Marginal Valuations of Waterfowl

Equation form and method of valuation	$\frac{V_{(D/Z)}}{Z}$	$\frac{V_Z}{D/Z}$	V_D
Equation (3-1), $\ln V = {}'f[\ln(Y, S, E, D/Z, Z)]$			
Mean of calculation for each individual over entire sample of 1,511	\$3.35	\$5.21	
Calculation at mean values of the untransformed independent variables	2.38	3.70	
Equation (3-2), $\ln V = {}''f[\ln(Y, S, E, D)]$			
Mean of calculations for each individual over entire sample of 1,511			\$4.37
Calculation at mean values of the untransformed independent variables			3.29
Equation (3-6), $\ln V = {}'''f[\ln(Y, S, D)]$			
Calculation at mean values of the untransformed independent variables			3.08

$$V_D = q_4 q_0 Y^{q_1} S^{q_2} E^{q_3} D^{q_4 - 1} \tag{3-5}$$

In a similar fashion, although somewhat less directly, the two marginal valuations of waterfowl developed in chapter 2,

$$\frac{V_{(D/Z)}}{Z} \text{ and } \frac{V_Z}{D/Z}$$

may be taken using equation (3-1). Results of the marginal calculations are shown in table 2.

Each of the marginal valuations has been calculated for every individual in the sample of 1,511, and the means of these marginal valuations are reported for both equations (3-1) and (3-2). The technique allowed a

---

formed state has been determined, rather than a conditional mean function, as has been shown by Arthur S. Goldberger in his article, "The Interpretation and Estimation of Cobb–Douglas Functions," *Econometrica*, vol. 35 (July 1968), pp. 464–472. Let us take the simplest case where a multiplicative relationship is assumed between a variable Y and a variable X:

$$Y = b_0 X^{b_1}$$

Parameter estimation is attempted by a least-square regression of the logarithmic transformation. From standard assumptions, the observed values of $\ln Y$ are normally distributed at each value of $\ln X$; the mean and median of a normal distribution are of course equal. But from the lognormal assumption it follows that in the untransformed state the distribution of Y values at each X value is skewed toward the higher values of Y, and the mean is greater than the median. (Both functions are approximations, as both estimators of b_0 are biased.) From equation (3-5) it is obvious that the value of b_0 enters into the value of the partial derivative and hence into the marginal valuation of a waterfowl. Since from equation (3-2), $s^2 = 1.71$, the marginal valuation estimated from the conditional mean function would be approximately 2.3 times that estimated from the conditional median function. As the relatively high value of s^2 and the relatively low value of $\bar{R}^2$ are closely related, there is reluctance to adopt the conditional mean function, and the conditional median function therefore is accepted.

in other geographic areas. According to a 1963 U.S. Bureau of Sport Fisheries and Wildlife survey, the percentage of individuals who paid to hunt waterfowl on private property was indeed small.[22] For instance, on the Pacific flyway (which includes those states and portions of states lying west of the Continental Divide) 3.1 percent of the hunters used such land, while hunting on ground owned or leased by private clubs was reported by 8.1 percent and 8.5 percent respectively of the survey respondents. The figures for California were considerably higher with respect to hunting on owned or leased private clubs: 16.3 percent reported hunting in each of the two categories. In any event, these persons represent a minority of California hunters and generalization to the population seems unwarranted. Approximately 26 percent of California hunters used public hunting areas for which there was an access fee,[23] but the rather elaborate advance reservation system used by the state clearly shows that the fee was not a market-clearing price.

Another possibility is to utilize the travel–transfer cost method to obtain incremental values for waterfowl. If two hunting sites differ in the availability of waterfowl for a season, but are virtually identical in all other respects,[24] and if it is possible to adjust for such hunter characteristics affecting use as incomes, tastes, and availability of substitute recreational pursuits, and if an adequate formulation of the cost method is obtained, a measurement of the value of the additional waterfowl available at the better hunting site is feasible. For the valuation to have validity over a wide geographic area, the experiment would have to be repeated at a fairly large number of paired sites, an expensive process in terms of both money and time. Given this fact and the authors' obvious uneasiness regarding the accuracy of the cost methods, the idea represents a road not taken.

A remaining avenue, on which we do journey, is represented by a modification of the Davis technique. We believe the scope of valuation questions may be limited in such a way that subsidiary satisfactions of the recreation day (to repeat, such as exercise and exposure to natural surroundings) are effectively excluded. This is done by confronting the recreationist with the possibility of doing without waterfowl hunting of a

[22] Robert G. Heath and M. Edwin Rosasco, "National Survey of Waterfowl Hunting on Various Types of Land Ownership," Administrative Report No. 25 (Laurel, Md.: Migratory Bird Populations Station, U.S. Bureau of Sport Fisheries and Wildlife, 1963).

[23] Since some individuals hunted on land representing different types of ownership, total percentages for all types add up to 116.9 in California. Heath and Rosasco, "National Survey of Waterfowl Hunting," table 4.

[24] This allows imputation to waterfowl of value differences, as other components giving value to the recreation experience are presumably equally available at both sites.

Valuing a Recreational Component

All the valuation methods so far discussed deal with measurement of net benefits for the total recreational experience, and in most cases, with attribution of those benefits to the value of the site. But where valuation of waterfowl (or other game, or fish) is desired, two additional complexities arise. The first is that the bagged waterfowl are only one component among many that contribute to the value of the recreation experience, and if other components give value, such as exercise and exposure to natural surroundings, then not all the value of the recreation may be attributed to the bagged game; some segregation of values must be devised to include all the characteristics of the recreational experience. When the value properly attributed to the birds is obtained, it presumably will be a measure of total net benefits for all of the waterfowl in a specified geographic area, as all the valuation methods discussed produce *total* net benefit figures. Such a figure would be of interest if all of the birds in the area were possibly to be destroyed by the actions of man, but it is far more likely that man's actions will cause incremental or decremental changes. Hence the second complication is to determine marginal net benefits from waterfowl rather than total net benefits.

To simulate as nearly as possible a market-determined purchase price, a marginal value of bagged waterfowl to hunters is indeed required, so let us examine available alternatives.

Jon H. Goldstein attempted to quantify the marginal net value of a duck through a regression technique in which the dependent variable was the rental payment of a hunter for seasonal hunting rights on privately owned property, and the independent variables were travel distance from home to hunting site, income, and ducks shot per season on the land.[20] The first derivative of rental value with respect to the number of ducks shot was the estimated marginal net value of a duck. Data for the regression were obtained from responses to a questionnaire mailed to Minnesota waterfowl hunters. As Goldstein himself acknowledges, "the model proved inappropriate,"[21] for less than 10 percent of the sample indicated they rented waterfowl hunting lands. Extrapolation from a very select group to the entire Minnesota waterfowl hunter population indeed does seem inappropriate.

Goldstein's formulation appears to have little more chance of success

[20] *An Economic Analysis of the Wetlands Problem in Minnesota*, Ph.D. thesis, University of Minnesota (Ann Arbor: University Microfilms, 1968). See also Jon H. Goldstein, *Competition for Wetlands in the Midwest* (Baltimore: Johns Hopkins Press, for Resources for the Future, 1971).

[21] Goldstein, *An Economic Analysis*, pp. 65–66.